PRAISE FOR
Three Days at Camp David

"This is a terrific book, a distant mirror on the present."
—Tom Friedman, three-time Pulitzer Prize–winning journalist

"There are at least three reasons to read *Three Days at Camp David*. It is a fascinating character-driven story. It provides critical historical perspective on America's role in the world, including the constant tension between nationalism and global engagement. And it sheds light on President Biden's challenge to restore America's alliances just as Nixon and Kissinger did in Garten's account of the early 1970s."
—Walter Isaacson, author of *Kissinger, Steve Jobs*,
and *Leonardo Da Vinci*

"Jeffrey Garten's *Three Days at Camp David* is a riveting account of one of the most consequential—if often overlooked—moments in financial history. Garten gives us Richard Nixon before the disastrous fall, making the pathos of the events all the more poignant. It's required reading for anyone eager to understand the world's current economic challenges."
—William D. Cohan, author of *The Last Tycoons: The Secret History
of Lazard Frères* and *House of Cards: A Tale of Hubris
and Wretched Excess on Wall Street*

"Garten tells a tale of gold and guarantees, secrecy and surprise, and conflict and cooperation. His fast-paced history brings the characters alive, explains complexity with clarity, and reveals America's competing impulses of leadership and retrenchment—a tension alive today."
—Robert B. Zoellick, former president of the World Bank, former
US trade representative, and author of *America in the World:
A History of U.S. Diplomacy and Foreign Policy*

"As a senior economic staff member working for National Security Advisor Henry Kissinger at the time, I was involved in many of the pivotal events in this book. Jeffrey Garten has told the story of how the United States severed the link between the dollar and gold in all its economic and foreign policy dimensions, and in all its human drama. Not only has he made a contribution to understanding this critical series of events in American history, but he has also pointed to the profound implications for America and the global economy in the years ahead."
—Robert D. Hormats, former senior White House and State Department official in five presidential administrations, and former vice chairman of Goldman Sachs (International)

"A fascinating play-by-play account of one of the most important events in the history of the modern world economy, with powerful implications for current policy."
—C. Fred Bergsten, founding director of the Peterson Institute for International Economics, and former assistant secretary of the Treasury

"A gripping tale of a critical period in global finance with striking relevance to our present moment, when the global system is in transition and so many pressures exist for America to look only inward."
—Merit E. Janow, dean, School of International and Public Affairs, Columbia University

"Jeffrey Garten makes a strong case that the end of the dollar's convertibility into gold has deep, long-run implications for the United States and the global economy. He provides valuable food for thought for our own uncertain era."
—Robert E. Rubin, former US Treasury secretary

"If you ever wanted a ringside seat to see how a world-changing decision is made in Washington, read *Three Days at Camp David*. It reads like a novel, yet it is a work of history, politics, and economic policy."
—Susan C. Schwab, professor emerita, University of Maryland, former US trade representative

"An enlightening study of an era when previously unthinkable economic measures suddenly went mainstream. . . . President Richard Nixon's high-wire economic policy of monetary anarchy, tariffs, and government wage-and-price controls is dissected in this incisive history. . . . Garten vividly sketches the personalities behind the policy. . . . [His] lucid, easy-to-grasp exposition focuses on international turmoil in exchange rates and trade."   —*Publishers Weekly*

"Highly charged account of the Nixon administration's abandonment of the gold standard. . . . Garten delivers incisive portraits of key players. . . . Fiscal and monetary policy wonks will admire Garten's skillful narrative and thorough research."   —*Kirkus Reviews*

"Important. . . . The challenges faced by the Nixon administration—from changing views about America's place in the world, through the pressures of globalization, to the difficulties in balancing economic and foreign policy—are exactly those facing President Joe Biden today. . . . Garten is uniquely well positioned to tell this tale."
—Rana Foroohar, *Financial Times*

"This is a terrific, suspenseful read. It is part of a sophisticated yet understandable explanation of what went on at a crucial moment in US economic history. But it also reads like a Hollywood script."
—David Smick, editor of *The International Economy*

"A great read about an historical event, and it's just as relevant fifty years on."   —Scott Horsley, chief economic correspondent, NPR

"The book does a fantastic job of drawing attention to a pivotal moment in history concerning the shifting balance of political power and the changing nature of money."
—Gillian Tett, chair of the editorial board, and
US editor at large, *Financial Times*

"A fascinating . . . case study." —Justin Fox, *New York Times*

"It is as readable as a novel, yet says a lot about the world we now inhabit."
—John Authers, *Bloomberg Opinion*

"Terrific new book." —Charles Lane, *Washington Post*

"It would be trite to say that Garten's book belongs on every international economist's bookshelf. It doesn't. It belongs on their bedside tables as light, but thoroughly enjoyable, reading and a useful reminder that, whatever economic forces might be at play, it is people, personalities, and politics that make history."
—Atish Rex Ghosh, IMF historian, *Finance and Development*

"A fascinating account. . . . [I] strongly recommend it to all enthusiasts of economic history and political economy."
—Alan Murray, CEO, Fortune Media

"[*Three Days at Camp David*] offers a clear presentation of the often arcane and obscure monetary issues involved, vivid portraits of the government officials who took part in the deliberations leading up to the Nixon announcement, and a useful and in some ways surprising assessment of the August 15 initiatives from the perspective of fifty years." —Michael Mandelbaum, *American Purpose*

"[A] gripping account." —Niall Ferguson, *Bloomberg Opinion*

"A richly detailed, character-driven account."
—Barry Eichengreen, *Foreign Affairs*

"Jeffrey E. Garten captures [Nixon's] decision-making in high dramatic style, turning what could have been a dry primer on international economics into a brisk adventure." —*Air Mail*

"In this outstanding book, Garten explains in detail what happened and draws lessons for today." —Martin Wolf, *Financial Times*

THREE DAYS
AT CAMP DAVID

THREE DAYS

AT CAMP DAVID

How a Secret Meeting in 1971 Transformed
the Global Economy

JEFFREY E. GARTEN

HARPER

NEW YORK • LONDON • TORONTO • SYDNEY

FOR INA,
WHO LIGHTS UP MY LIFE EVERY SINGLE DAY

CONTENTS

III. THE WEEKEND

IV. THE FINALE

Introduction

At 2:29 p.m. on Friday, August 13, 1971, President Richard Nixon walked from the West Wing of the White House to the South Lawn, where he boarded Marine One, the presidential helicopter. A young marine officer crisply saluted the president as he ascended the six steps to the cabin door. The rotary blades atop the fuselage slowly churned horizontally while those on the tail wing began to spin vertically. After a seemingly long moment, Marine One then slowly lifted straight up. Within seconds, its nose dropped, and the aircraft banked to the northwest, flying past the Washington Monument and a landscape of urban buildings. It then flew over small towns and continued over the thickly forested eastern Blue Ridge Mountains on its thirty-minute journey.

Aboard the helicopter, Nixon was crammed in with several men—Treasury Secretary John Connally, Federal Reserve Chairman Arthur Burns, Chairman of the Council of Economic Advisers Paul McCracken, Director of the Office of Management and Budget George Shultz, and White House Chief of Staff H. R. Haldeman. Also on board were two pilots and a flight engineer, two Secret Service agents, and one military aide who carried a suitcase with the nuclear codes.

The seats were upholstered in gold cloth and rested on a rich blue carpet. Nixon had seated himself in an armchair on the right side of the aircraft. With his back to the cockpit, he could look out on the rest of the cabin. A telephone with a long, coiled extension cord was affixed low on the wall, at knee level—the only source of communication between the passenger cabin and the world outside. The space was cramped; briefcases would have to rest on passengers' laps, and a person could barely move down the aisle unless he walked sideways.

Another helicopter had taken off from Andrews Air Force Base in Maryland one hour earlier. On board were Paul Volcker, undersecretary of the treasury for monetary affairs; Peter Peterson, assistant to the president for international economic affairs; Herbert Stein, a key member of the Council of Economic Advisers; William Safire, a White House speechwriter; and John Ehrlichman, director of the White House Domestic Council. Several staffers had departed even earlier by car for the ninety-minute trip.

All these men were headed to a top secret meeting at Camp David, the presidential retreat in Maryland's Catoctin Mountain Park. When Haldeman and his deputy, Lawrence Higby, contacted the twelve invitees the prior afternoon, each was told to pack an overnight bag, get ready to leave town for a few days, and under no circumstances tell their families where they were going. In the end, there would be fifteen attendees, including Nixon, Haldeman, and Michael Bradfield, a Treasury lawyer who joined the group on Saturday morning. Only a few participants knew the specific agenda, but most understood that the meeting would be momentous. The day before, George Shultz, the powerful director of the Office of Management and Budget, described his view of the upcoming event to Nixon, saying, "This is the biggest step in economic policy since the end of World War II." On the way to their helicopter that very morning, Herbert Stein told William Safire, "This could be the most important weekend in the history of economics since Saturday, March 4, 1933" (the day President Franklin D. Roosevelt closed all the banks in America). In fact, the group was about to ignite a series of explosions that would rock America's political alliances, set the U.S. dollar on a radically new course, and reshape the U.S. and global economies.

THIS BOOK TELLS the full story of that weekend—what led up to it, what transpired over those three days, what happened afterward, and what it all meant. It is, above all, an account of how the United States began to rethink its role in the world, how it realized it could no longer shoulder all the burdens that were thrust on it after World War II, and how it attempted to redistribute part of its awesome responsibilities to its allies. The events I have written about also illustrate a major inflection point in American history: a moment when the United States, its enormous power notwithstanding, was forced to recognize its growing interdependence with other countries and the need to move from using its raw unilateral leverage to engaging in more multilateral diplomacy and cooperation.

I have told this story through the lens of the U.S. dollar and its relationship to gold. The issues surrounding the American currency may sound esoteric, but in fact the dollar was, and still is, at the intersection of foreign policy, national security, and international commerce. Our currency is also a major influence on what we care so passionately about at home, including such issues as jobs, prices, retirement security, financial stability, economic opportunity, and even respect in the world. We may not think about the dollar in such all-encompassing terms, but the value of the greenback has had a broad and deep impact on the country and the world as well as on our everyday lives.

A BRIEF HISTORY helps to understand the immense challenge facing Nixon and his team as they were traveling to Camp David. At the time of the meeting, the United States had a longstanding treaty commitment, made in the context of joining the International Monetary Fund (IMF) in 1944, that any foreign government or central bank holding U.S. dollars could exchange them for gold. The actual rate of the exchange—$35 per ounce—had been set by President Franklin Roosevelt in 1934, and it had never been changed. In addition, other currencies, such as the British pound, the West German mark, or the Japanese yen, were linked to the dollar at a fixed exchange rate. For example, in 1949 one dollar was worth 360 yen, and the ratio could not vary more than 1 percent either way. That dollar–yen rate could be changed more than 1 percent only as a last resort in a long-term emergency. All these provisions were contained in the IMF Articles of Agreement.

These obligations stemmed from a global monetary conference in 1944 at Bretton Woods, New Hampshire. It was there that the wartime allies, principally the United States and Great Britain, created a new international monetary framework for the post–World War II era. The underlying idea was that a global financial system—with exchange rates that couldn't fluctuate against one another beyond 1 percent, and with rates that were tied ultimately to gold—would provide the stablest possible background for countries to sell one another their grains, food, machinery, autos, textiles, and other products.

The new monetary setup was designed to create a foundation for a world that was much different from the one that existed in the turbulent 1930s. In those earlier years, although countries pegged their currencies to gold, the gold standard was both much more rigid than the Bretton Woods arrangements, and much less internationally supervised. As a result, countries participated inconsistently, and eventually abandoned the system altogether. And once the system collapsed—because of vicious, unfair competition combined with the economic slowdown of the Depression—governments erected barriers to trade, such as tariffs and quotas, that distorted the natural flows of trade and money. They also pushed down their exchange rates to reduce the prices of their exports—what economists call "competitive depreciations."

Here is a simple, hypothetical example of how that worked, and of the pernicious effect on world trade and economic growth that followed: Suppose the German mark had been linked to the French franc at a ratio of one mark to two francs. Assume also that the price of a ton of German steel had cost 10,000 German marks, while the French charged 20,000 francs for the same tonnage. If Germany had devalued its exchange rate by 10 percent against gold and other currencies, it would have essentially discounted their price by 10 percent relative to the French price. Not standing for that, Paris would have proceeded to devalue its currency to underprice the German metal, perhaps pushing down the franc so that a ton of steel cost 16,000 francs—a 20 percent depreciation amounting to twice the discount offered by Germany. And so it would have gone, with Belgium and Italy also devaluing their currencies even more than the French did, in order to gain competitive advantage. As Germany's and France's cheaper exports penetrated the U.S. market, America would have resisted the flood of incoming products by raising its tariffs or by establishing quotas. As imports into the United States contracted, exports to the United States by Germany, France, and other U.S. trading partners would likewise have decreased. Other importers of steel would have reacted the same way the United States did.

Indeed, during the reign of the gold standard, competition like that in my illustrative scenario shrank the value and volume of trade for everyone, which in turn led to a downward spiral of global economic growth. The extreme trade protectionism that resulted made the Great Depression worse and created strains that likely contributed to the onset of World War II.

Mindful of such disastrous experiences, Bretton Woods was built around a new kind of gold standard. The U.S. dollar would be at the center of the system. Washington agreed to value one ounce of gold at $35. Every other currency was valued at a fixed rate to the dollar. Being the fixed point, the dollar could not be devalued. Any government or central bank could redeem the dollars they held by going to the "gold window" at the U.S. Treasury. The so-called dollar–gold system, with all other currencies being fixed to the greenback, was designed to create an environment that would provide stability and predictability to international trade.

But that's not all. The leaders at Bretton Woods wanted to create a web of rules and obligations that would prevent the protectionism of the 1930s. A core idea was to provide countries that were running large trade deficits with more flexibility to get their economies back into balance. Before Bretton Woods, nations with big trade deficits had few alternatives other than to erect barriers to imports or to depreciate their currencies. In both cases, doing so would not only disrupt free-flowing trade but also create an environment in which other nations would retaliate, leading to slower growth and fewer jobs. Under Bretton Woods, governments running trade deficits had more options.

First, the new rules allowed some flexibility to adjust exchange rates up or down by 1 percent. Unlike the rigidity of the 1930s gold standard, this was an automatic right and would not set off alarms. Second, a new organization called the International Monetary Fund was established to oversee the new financial system and to provide substantial loans to countries so that they had time to change their economic policies without having to raise trade barriers or depreciate their currencies. Third, if their trade deficits were due to structural problems that were truly severe and long term, countries could devalue or revalue their currencies under the auspices of IMF rules and oversight. This would be done according to a process that other governments understood and, via their membership on the IMF Board of Governors, approved. Retaliation via competitive devaluations was not allowed. In other words, the Bretton Woods system was built around an international institution and a body of international law and understandings, none of which existed in the 1930s.

To be sure, a major contradiction was inadvertently built into the new dollar-centered global monetary system. The underlying assumption in the 1940s and early 1950s was that as the world economy recovered from the war, the economies of Western Europe and Japan would resume their growth and that international trade would expand accordingly. This revival would require more capital, much of which would come from the United States in the form of dollars. However, the more dollars that circulated, the more the law of supply and demand would cause each greenback to be worth less to those who used it. Over time, therefore, a fundamental readjustment of the Bretton Woods monetary system, based on the centrality of the dollar, would inevitably be required. In other words, the postwar monetary arrangement contained the seeds of its own demise.

But even those officials who understood the inevitability of having to create a new monetary order down the road saw no alternative to protecting the system in the 1960s. In fact, the dollar–gold link was enthusiastically supported time and again in public statements by Presidents Kennedy and Johnson. Both leaders saw a strong, stable greenback as one critical element of America's leadership of the free world. On Thursday, July 18, 1963, JFK told Congress, "I want to make it . . . clear that this Nation will maintain the dollar as good as gold, freely interchangeable with gold at $35 an ounce, the foundation stone of the free world's trade and payments position." On Wednesday, February 10, 1965, President Johnson made a similar pledge to Congress: "The dollar is, and will remain, as good as gold, freely convertible at $35 an ounce." It wasn't just the two Democratic presidents who espoused the policy, either. In both administrations, their secretaries of treasury and the chairman of the Federal Reserve (William McChesney Martin in both cases) left no doubt that the United States was committed to the dollar–gold link.

Although President Nixon was more circumspect than his predecessors in his first two and a half years, he continually emphasized America's determination to follow policies that would keep the dollar strong, and he never denied America's commitment to maintain the dollar–gold link. In addition, during this time, his two successive secretaries of the treasury, David Kennedy and John Connally, as well as his Federal Reserve chairman, Arthur Burns, explicitly supported the commitment to exchange the dollar into gold at $35 an ounce, if foreigners requested the conversion.

Thus, up to midsummer 1971, in the eyes of foreign officials and traders and investors, the $35-an-ounce dollar–gold link was seen as being unconditionally backed by the U.S. government. This perception gave other nations confidence to keep accumulating dollars with the assurance they would always be able to cash them in for a tangible alternative asset—i.e., gold.

THE TWENTY-FIVE YEARS following the war produced a massive recovery of war-torn Western Europe and Japan, and two decades of unprecedented prosperity in the non-Communist world. The stability resulting from the dollar–gold link was a central element of that progress. Altering that link carried unknown risks of subverting what was an extremely successful financial system at the heart of the broader international economy. That system was also a critical underpinning of America's political and military alliances in the heat of the Cold War. After all, expanding trade and investment strengthened America's partners and made them more self-confident. It also tightened the ties that bound the free world together, not just economic links but philosophical support for the combination of free markets and democratic political systems.

Well before the Camp David meeting, however, it had become clear to Americans and foreigners alike that the United States had nowhere near enough gold reserves to make good on its commitments. Uncle Sam had been flooding the world with dollars through the Marshall Plan, through other foreign aid programs, through its financing of U.S. troops and bases abroad, and through the growing foreign investments of U.S. multinational companies. But all the while the supply of gold was not increasing at nearly the same rate as the dollar outflow. And even when gold supplies were expanding, Europe and Japan were accumulating their share of it, leaving less for the United States. Thus, a major gap emerged between America's gold reserves and official holdings of dollars held abroad—dollars held by central banks and governments that were eligible to be convertible into the precious metal.

The gold drain relative to the dollars circulating outside the United States was dramatic. In 1955, the United States had enough gold—$21.7 billion worth at the $35 price—to cover its liabilities to other central banks and governments, which totaled

$13.5 billion. In other words, American gold reserves exceeded official dollars abroad by over 160 percent. By the summer of 1971, however, America had just $10.2 billion worth of gold, compared to official foreign dollar holdings of $40 billion. Thus, Washington had only 25 percent of what it needed to make good on its commitment to exchange gold for dollars.

Although foreigners had been worried about the gold drain for several years, they simply didn't want to come to grips with the imminent possibility that they would ask to exchange dollars for gold and be rebuffed by an Uncle Sam, who simply didn't have enough gold to make good on his promise. The dollar–gold problem seemed too big and too complex, and no one was sure how to fix it without causing major global upheavals. The very act of America's withholding gold from holders of dollars could also have brought down the global economy, as it would have shattered the assumption held by traders, investors, and business-men that the dollar–gold link was the foundation for a stable and predictable global market.

AT 9:00 P.M. on Sunday, August 15, 1971, after two days of intense discussions at Camp David, and with just a few hours' prior notification to other governments, Nixon unilaterally sev-ered America's long-standing commitment to the dollar–gold link at $35 an ounce. Put another way, he closed the "gold window" at the U.S. Treasury. The Bank of England, the West German Bundesbank, the Bank of Japan, and their central bank counter-parts were left with hundreds of millions of dollar reserves that, after a quarter century, had suddenly lost their gold backing and were therefore of uncertain value.

In announcing this fundamental change in what was consid-ered a sacred U.S. obligation, Nixon pulled a central plank out of the edifice of international finance and of the regime of international trade that depended on it. He shook to the core U.S. relationships with Western Europe and Japan that U.S. administrations had spent the postwar era building, alliances critical to the ongoing, all-consuming rivalry with the Soviet Union. The fallout was extensive, and a series of politically contentious follow-up negotiations ensued over the next few years.

On that Sunday, August 15, 1971, Nixon was not just announcing a change in monetary policy. He was in essence telling the world that the near-omnipotent role that the United States had played since the war was over. The days in which America shepherded Western Europe's and Japan's recovery, the era in which it had opened its markets to imports without receiving reciprocal treatment, the years in which it funded a disproportionate amount of the common military defense, the quarter century in which it held up the monetary system with its gold—all that was now going to change. Washington was not just asking its allies to enter a new age of burden-sharing; it was forcing them to accept it. The United States was also taking the first steps toward ushering in an international system in which it would have to pursue multilateral, as opposed to unilateral, policies. This turn toward multilateralism wasn't necessarily Washington's enthusiastic preference, but it was a bow to the reality of shifting power in the world.

THIS AUGUST WEEKEND was therefore a watershed in modern American history. It represents a tale of the importance of the dollar not just as a way to facilitate trade, or as a currency one could confidently invest in for the future, but also as an instrument of changing American power and influence. It is an account of how the world's most powerful nation made a decision to fundamentally change a quarter century of one of its core policies, forcing the rest of the world to adjust to the impact of its disruptive action. It is the story of how some of America's most talented, knowledgeable, and experienced public servants wrestled with a daunting set of decisions affecting the United States and the world for generations afterward. That effort was not devoid of strongly opposing views, personal animosities, or bureaucratic infighting, but in the end, it was also characterized by impressive teamwork.

I WANTED TO tell the story of this August weekend at Camp David for a number of reasons. First, what happened over those three days parallels a number of issues confronting the United States today, and I found that exploring the history of events that took place a half century ago sheds light on our current challenges. Then, as now, the United States was asking itself the most basic questions about its place in the world. By the late 1960s and early 1970s, prominent politicians and ordinary citizens alike were convincing themselves that America gave much more than it received from the burdens of its leadership role. Then, as now, America was pushing for a trading system it deemed to be fairer to the United States. Then, as now, Washington was unhappy that its responsibilities for maintaining political alliances imposed too heavy an economic cost on it. In Nixon's time and in ours, therefore, Washington was pressuring its NATO allies to contribute more to the common military defense. Then, as now, the nation was unsure of how to respond to increasing globalization, including growing trade imbalances, job-competing imports, and the outsourcing of jobs to foreign countries by multinational corporations—all in their early stages in the 1970s but nevertheless already contentious. In the early 1970s, the Federal Reserve was under an intense public spotlight as it struggled to deal with the new economic conditions. On that score, nothing has changed. Then, as now—Vietnam then, Afghanistan today—a long war was ending without any semblance of American victory and was sapping the country's appetite for military involvement overseas. A half century ago, Congress and the public were demanding more attention to domestic needs, such as infrastructure, education, civil rights, and an adequate social safety net—same as today. In Nixon's time, America's allies worried Washington would turn inward, becoming nationalistic and protectionist. In recent years, their fears have been coming true. In the late 1960s and early '70s, the country was beset by partisan politics, although that situation is far worse today. In 1971, many said the dollar was "overvalued," causing imports to be too cheap and exports to be too expensive, the same charge frequently made over the past few years. In 1971, Nixon recognized that other nations, particularly Germany and Japan, should shoulder more responsibility for the management of the international political and economic order. The only thing that has changed is that today the United States needs the help of a broader range of nations to achieve all its goals. Germany and Japan are still among them, but now the group also includes China and a host of other up-and-coming countries such as India, Brazil, and South Korea. In Nixon's time, a crying need for major international monetary reform existed, including new rules for currencies. When the coronavirus pandemic is behind us, with widespread human and physical destruction in its wake, and with national debts at wartime levels, we may very well face requirements for new global financial arrangements that match in scope the extensive reforms made at Bretton Woods.

But there are also big differences between the early 1970s and now. It might appear that the Nixon administration was taking the initial steps of the America-first policy that has returned in acute form in the years of President Trump. But as we shall see, what began in 1971 as a harsh, unilateral set of actions to sever the dollar–gold relationship actually had the effect of increasing America's involvement in the global economy, expanding its investment in international organizations, and deepening international coordination between the United States and its allies. Even though Washington had delivered a severe shock to Western Europe and Japan, Nixon never contemplated abandoning the principle of working closely with America's partners to deal with problems through consultation on nuclear arms treaties, global poverty, global food security, and explosive oil prices, to cite just a few examples. In 1971, Washington never eschewed the idea that more rather than less trade was good. It continued to search for a better system for managing currencies. It never lost sight of the fact that, over time, economic and political ties became intertwined, and when they moved in the right direction, they strengthened democratic societies. This mind-set was due in large part to the background, talents, and world outlook of Nixon and most of the advisors around him. The result was that the pattern of cooperation among the United States, Western Europe, and Japan lasted for over four decades. In that respect, this book tells the story of how Washington forced big changes in the international arena while preserving the underlying political framework that is designed to enhance and not destroy the benefits of nations working together to solve big problems.

With a new administration now taking hold in Washington in 2021, a comparison with the situation a half-century ago is illuminating. At Camp David in 1971, Nixon embarked on a tough, even confrontational approach with America's closest allies, but having won their undivided attention, Washington pivoted to a policy of decades of international cooperation. The big question now is whether the Biden administration can use the "Trump shocks"—the defiant unilateralism; the withdrawal from international agreements on climate change, and from international institutions such as the World Health Organization; and the rampant use of tariffs and economic sanctions, all of which certainly caused deep concerns among America's allies—to re-engage them in serious negotiations about rebuilding a new cooperative world order appropriate to the new challenges ahead.

I was also drawn to write about that Camp David weekend because it was a world-changing event with ramifications visible today that have been neglected by historians. You can find magazine and academic articles on the subject, or references in the memoirs of some of the Camp David participants and in histories of international finance. But I have not seen a book interpreting the weekend itself for a general audience, and certainly nothing that tells the story with the heavy emphasis I have given to the participants who shaped and made the decisions.

The events I have written about here also resonate with my personal experience. During my professional life, including time spent in economic and foreign policy positions in the Nixon, Ford, Carter, and Clinton administrations, my many years on Wall Street, and my experience at the Yale School of Management as a dean and professor, I have been able to interact with several of the prominent personalities in this book, including Paul Volcker, Henry Kissinger, and Peter Peterson. I have also known and worked with a number of senior staff who participated in the Camp David weekend and with many outside experts who observed what happened during the weekend and afterward. It has been my good fortune to have interviewed a large number of these people for this book. (See page 421 for the list of interviews.)

YOU DON'T NEED to know much about economics, finance, and trade to read and understand this story. I am writing about the dollar not from a technical point of view but through the lens of history, people, and politics. Therefore, I have simplified some issues and left out details that would be found in more in-depth economic analyses, in order to explain what happened at Camp David to an audience with little or no background in economics or finance.

I begin with the explosive situation facing Richard Nixon after his first two years in office, before turning to the key members of his team who were central to the weekend and its aftermath. (See page 333 for a complete list of the key figures mentioned in this story.) It is through the lives, thoughts, and actions of these men that I describe the experience, ideas, biases, and character traits each brought to the table at the August 1971 weekend gathering. After a detailed account of the days leading up to Camp David, I turn to each of the three days of the weekend itself. Ordinarily, the end of a story might be the point where decisions have been made, but in this case, it was the way the decisions were executed that constitutes a major part of this saga. Thus, I discuss what actually happened in the weeks, months, and years after the Camp David event and assess the longer-term impact. I conclude with my view of how to think about the weekend a half century later, including what lessons we can draw for the future.

IN HIS ACCLAIMED history of the Federal Reserve titled *Secrets of the Temple*, journalist and author William Greider wrote, "If historians searched for the precise date on which America's singular dominance of the world's economy ended, they might settle on August 15, 1971." He was right on the mark. Let me explain why.

I.

CURTAIN UP

Richard Nixon Ascending

Richard Nixon assumed office in 1969 having defeated his democratic rival, Vice President Hubert Humphrey, by just 500,000 votes out of 73 million cast and by a hairline margin in the Electoral College. The elections also produced a Congress in which both chambers were controlled by Democrats, the first time since 1848 that a new president was confronted by a legislature in which the opposition was in charge. "Mr. Nixon starts with no clear mandate from the people, no great fund of personal popularity, and an opposition Congress that contains many elders of both houses who have regarded him with suspicion and even personal hostility ever since he was in the House of Representatives a generation ago," wrote James Reston of the *New York Times*.

Among the reasons for so close an election were dramatically contrasting conditions in the country. On the one hand, the Johnson years were ones of extraordinary prosperity, including low unemployment and the absence of recessions. On the other hand, the nation seemed to be unraveling. In 1968 alone, America was torn apart by the Vietnam War, urban riots, and violent student activism. It was the year of the assassinations of Martin Luther King Jr. and Robert F. Kennedy, and a Democratic National Convention in Chicago notable for its violent confrontations between police and demonstrators in the surrounding streets. Shortly after the inauguration, in fact, Warren Christopher, who had stepped down as deputy attorney general for LBJ, came to the White House to see his old classmate from Stanford Law School, John Ehrlichman, who had just been appointed as a senior aide to the new president. Christopher handed Ehrlichman a package of documents. "They were proclamations to be filled in," Ehrlichman told Nixon biographer Richard Reeves. "You could fill in the name of the city and the date and the President would sign it and declare martial law."

When Nixon came to power, America had been fighting for the past several years to defend South Vietnam from being taken over by the Communist regime of North Vietnam, which itself was supported by China and the Soviet Union. America's goal was to make sure South Vietnam and its neighbors in Southeast Asia did not fall into the Communist camp. Like his predecessor, Nixon was determined not to be the first American president to lose a major war, and in his campaign, he had pledged to end the fighting on terms deemed honorable for the United States. As of the date he took office, nearly forty thousand American soldiers had died in Vietnam (1956–68) and two hundred more were being killed each week. Public support for the war was eroding fast.

ON THE DOMESTIC front, Nixon had inherited an extensive array of domestic programs under the rubric of President Johnson's "Great Society"—an expansive view of the role of government not seen since the days of President Franklin Roosevelt's "New Deal." The overall program included the launching of Medicare, Medicaid, the broadening of civil rights, and programs to address poverty, poor access to education, and lack of economic opportunity.

Between the Vietnam War and the domestic programs, Nixon faced a painful policy dilemma. LBJ had encouraged Americans to believe the United States could mobilize its human and financial resources to fight a substantial war and, at the same time, vastly expand social programs at home. In other words, the country saw little truth in the classic concept taught in basic economics classes: the trade-off of guns versus butter. In fact, most Americans thought they could enjoy guns *and* butter. However, as the growing federal budget deficits and the rapidly deteriorating American trade position showed, Nixon was encountering a world that was imposing limits on America's national goals and on Washington's ability to afford them. The United States could no longer build and maintain a massive military machine and simultaneously extend the social safety net to all who needed it. The United States eventually would be forced to choose *between* guns and butter.

THE CHALLENGES NIXON faced in his first two years, 1969 and 1970, were acute. He began his ambitious plan to orchestrate an era of arms control agreements with the Soviet Union and the opening of relations between the United States and China. But he would not produce actual breakthroughs until 1971, and ending the Vietnam War remained the most pressing, and elusive, foreign policy goal. Moreover, relations with allies in Western Europe were becoming more difficult. Having recovered from World War II and wanting more independence from the heavy hand of Washington's dictates, America's overseas partners were restless. West Germany was seeking closer ties with East Germany, despite skepticism from a Cold War–obsessed Washington. France resented the dominance that the United States had achieved in the free world in both military and economic matters; it had expelled NATO headquarters from Paris, causing the organization to move to Brussels, and it deliberately undercut the dollar's role by demanding gold—far more frequently than other governments did—for the greenbacks it had amassed. Great Britain was less likely to follow American preferences than in the past because it was focused on joining the European Community (EC) and thus currying favor with France and West Germany, the EC's two most influential members. As countries such as West Germany and Japan became more economically competitive with America, the political tensions between them and Washington were growing. Controlling all these centrifugal forces was critical for the Nixon administration because anything less than a united allied front weakened its hand against the Soviet Union.

THROUGHOUT ITS FIRST two years, the administration was also combatting what was called the "New Isolationism," a strong negative congressional and public reaction to the tragic and unwinnable Vietnam War. The new isolationists wanted the United States to rebalance its resources away from foreign policy to domestic needs, including withdrawing from Vietnam and investing more in the inner cities, or spending less for NATO and providing more funds for food security to the poorest Americans. The pressure to look inward came from top congressional leaders like Senate Majority Leader Mike Mansfield (D-MT), who pressed the administration not just to end the Vietnam War, but also to bring back all troops stationed in Western Europe. Nixon and Kissinger took the threat from Mansfield and his colleagues with utmost seriousness. To them, dismantling American defenses abroad would have amounted to a humiliating retreat from America's commitments, not to mention a weakening of its own self-defense.

Congress also forced major reductions in the defense budget, causing intense battles with the administration. Kissinger, in his memoirs, wrote, "New military programs [were] fiercely attacked; some passed only by the thinnest margins; once they were authorized, their implementation was systematically whittled down and funds reduced annually." This tension between officials focused on national security and foreign policy on the one hand and those concerned with poverty, housing, education, and civil rights on the other was a constant feature of the times.

THE NEW ISOLATIONISM was all the more difficult to deal with given that global economic interdependence among nations was on the rise, with no end in sight. International trade, the movement of money across borders, increasing business and recreational travel, and the transmission of information around the world were ushering in a new level of globalization (although that word was not widely used until the 1980s).

By the 1970s, also, domestic jobs were being significantly affected by trade, with rising exports generating more jobs and rising imports undercutting them. High interest rates abroad caused dollars to stream out of the United States in order to earn higher returns. As a result, the Federal Reserve was forced to lure dollars back by raising American rates, an act that could slow down U.S. economic growth and make life miserable for millions of Americans.

It followed that for the first time since World War II, trade and international finance appeared at the center of America's traditional foreign relations agenda alongside more traditional issues such as control of nuclear weapons or the size and location of military bases abroad. The reason was that globalization brought with it a new set of domestic issues relating to employment. As the phenomenal rate of economic growth of the 1960s began to level off in the United States, job displacement caused by rising imports or by U.S. multinational companies moving part of their production overseas became explosive political issues. As Nixon took office, in fact, the efforts of Congress and labor unions to protect the economy from imports by raising tariffs and imposing quotas were increasing and becoming a potent political force. Such protectionism became intertwined with strong isolationist pressures stemming from aversion to the Vietnam debacle.

When it came to trade, Washington was particularly focused on several issues. The United States resented both the tariffs and quotas Europeans put on imports of agricultural products and the import preferences they gave to their former colonies and simultaneously denied to everyone else. Americans chafed at the way Japan blocked almost all imports through a maze of complex regulations, not to mention tacit collusion between Tokyo and its big corporate conglomerates to keep out U.S. products.

Sen. Abraham Ribicoff (D-CT), a key member of Congress on trade policy, captured the times in a report to his colleagues following a trip through Western Europe. He wrote that geo-economics was replacing geopolitics as the most important element of international affairs and that America was far behind its rivals. "While we concerned ourselves with the NATO order of battle," he wrote, "the Germans were more concerned over orders for Volkswagens."

NIXON KNEW THAT to sustain America's active leadership role in the world—in order to keep isolationism and protectionism at bay and respond effectively to intensifying globalization—he could not continue with the status quo. On a trip to Asia on Friday, July 25, 1969, he proclaimed that America would no longer provide troops to allies such as South Vietnam to fight internal or external wars. Washington could supply money and equipment, but not armies. The exception would be for countries with which the United States had defense treaties, such as those in NATO or Japan, or countries subject to a nuclear attack. The rest would have to supply their own troops. Called the "Nixon Doctrine," this represented the most dramatic change in U.S. foreign policy since the days of the Marshall Plan. It was the beginning of a major shift in America's role in the world, one that would soon extend to U.S. international economic and financial relationships. The new strategy would be highlighted time and again in Nixon's speeches and news conferences, and in comprehensive reports he would make to Congress on the administration's overall foreign policy.

The Economic Crisis

The pressures on foreign policy, the state of domestic unrest, the trade-offs between financing the Vietnam War and funding the Great Society programs, the growth of isolationism and protectionism, and the new focus on burden-sharing in defense spending—all were challenges enough for the Nixon administration. In addition, the economic problems at home seemed intractable. And with intensifying globalization, it was increasingly difficult to separate domestic and international economic problems.

All this proved worrisome to foreign holders of dollars, who believed that the United States was unable to get its own house in order when it came to controlling its inflation. Foreigners also focused on the growing balance-of-payments deficits, which showed the difference between how many dollars were flowing abroad versus how many were coming back to the United States. Large and continuing deficits had resulted in foreign governments and central banks, plus other holders of dollars, accumulating more greenbacks than they needed or wanted.

THE CONCEPT OF inflation is critical to this story. Inflation meant that prices would be constantly rising and that a dollar in the future would be able to purchase fewer products and services than it had been able to buy before the inflation began. Every year, the rising cost of a quart of milk, an automobile, a visit to the doctor, or college tuition would outstrip the rise in salaries and wages for most people. At the same time, the value of people's savings would erode, because inflation would diminish what they could actually earn on money they had put away for, say, their retirement. Thus, if the interest rate for a savings account was 6 percent and the rate of inflation was 5 percent, then savers would actually earn only 1 percent per annum.

Because inflation raised the prices of American products, it had a destructive impact on the balance of payments. In the first instance, American exports, such as machinery or grains, became more expensive and less competitive. Beyond that, inflated prices at home made imports, such as consumer electronics or clothing, cheaper and more attractive.

Inflation also meant that the dollar would be worth increasingly less in terms of what it could buy in world markets. This caused foreigners additional angst about holding on to the American currency, and it made the prospect of exchanging their dollars for gold more tempting. Many investors believed that the precious metal would hold its value over time much better than paper money. Whether that was true or not was less important than the widely held perception that it was.

The government had two conventional ways to quell inflation. The most direct and quickest route was for the Fed to raise interest rates to slow down the economy. The same goal could also be achieved by reducing federal spending or by raising taxes—in other words, by following a more restrictive fiscal policy. Put another way, the government had two dials to turn either alone or in tandem: monetary policy and fiscal policy. The prevailing theory among economists was that by raising interest rates or tightening the budget, the resulting slowdown in growth would reduce workers' demands for higher wages and also reduce the need for companies to raise their prices, both of which would translate into lower inflation. The big risk was that an engineering of slower growth could inadvertently go too far and push the country into a recession. Nixon's nightmare was that his administration would do just that.

RAISING INTEREST RATES and cutting the budget were also remedies to help America's trade balance by slowing the economy and reducing the nation's appetite for imports. However, the administration and Congress were not willing to take any fundamental monetary or fiscal measures for trade reasons alone. The rationale was that for the United States, total trade (exports plus imports) as a percentage of GDP was not significant, amounting to about 8 percent in 1970. The United States would be manipulating over 90 percent of the economy to influence the much smaller percentage that was exposed to trade. In effect, the tail (trade) would be wagging the dog (domestic economic growth and employment). Washington was not about to do that. This resulted in a situation in which other countries were worried about America's outflows of dollars and frustrated that Washington was engaged in what Western Europeans called "benign neglect." (For countries like West Germany and Japan, the need to take greater account of trade in their overall policies would have been different because trade as a percentage of their GDP was between 20 and 30 percent.)

BY 1970, INFLATION had reached 5.5 percent, having increased in each of the previous three years. Vast expenditures for the Vietnam War had been causing the economy to overheat, and the way to cool things down would have been to raise taxes and thereby reduce consumer spending. But for various political reasons, a tax increase wasn't enacted until the last six months of the Johnson administration. Moreover, higher taxes were proving ineffective in dampening the rising cost of living. Thus, inflation continued to undercut the prosperity that Americans had become accustomed to since the 1950s.

Nixon responded upon taking office with an economic program that he thought would reduce inflation and stabilize prices, all the while not increasing unemployment. His administration called the policy one of "gradualism." It had two components. Nixon would press the Federal Reserve to raise interest rates—but not too high or too fast—in order to slow the economy and thus reduce upward pressure on prices. In turn, the administration would not take dramatic action of slashing the federal budget, but instead would make small, constant, and very gradual spending reductions. Unlike orthodox remedies that would have involved much higher interest rates or severe budget cutbacks, gradualism promised very little pain or sacrifice. The administration believed that if it could demonstrate clear and credible intention of policy, the markets would get the message and ease up on pushing up prices.

By the end of 1970, however, it was clear that gradualism was a complete failure. Everything was going wrong. A recession was beckoning, while both inflation and unemployment were rising. Policy makers talked about a stubborn phenomenon, with which they had no previous experience, called "stagflation." They didn't know how to combat it, because whatever they tried backfired. If they slowed down the economy to fight inflation, they would worsen unemployment. If they expanded the economy to generate more jobs, they would accelerate inflation. To economic officials and private economists as well, stagflation became a problem for which they had no balanced solution. They would have to choose between objectives, less inflation or more jobs, but they couldn't have both.

Gradualism had not worked for several reasons. For starters, the Fed never believed in it and refused to keep money expanding at a steady rate. Also, budget deficits were not sufficiently reined in. In calendar year 1970, for example, Nixon's budget deficit was $11 billion, whereas the budget had been in surplus the year before by $9 billion—amounting to a $20 billion negative swing.

An equally critical impediment to gradualism was collective wage bargaining, the effort by the big labor unions to negotiate raises in pay and benefits with industry. In the early 1970s, the United States was then still a country built around manufacturing, and most of the biggest sectors, such as automobiles, heavy machinery, and the building trades, were unionized. Organizations such as the AFL-CIO, the United Auto Workers, and the International Association of Machinists and Aerospace Workers, exerted a powerful influence on national economic policy. Inflation forced labor to push up wages in its collective bargaining, which in turn caused companies to raise prices. For example, the unions would negotiate with an automobile company to win a raise of, say, 5 percent of the average worker's salary. Following that, the automobile industry would raise its prices by a similar amount. This would have nullified the raise achieved by union members, and so, in the next negotiation, labor would raise wages once again. This upward spiral of prices would continue, spreading from one industry to another. Nixon tried to rein in this unending sequence by cajoling and threatening unions and companies, but he was met with little success.

By the end of 1970, even Nixon's Council of Economic Advisers was brutally frank about where the economy stood. In its

February 1971 report, reflecting on the previous year, it said that the "performance of the economy disappointed many expectations and intentions, including those of the Council . . . The [GNP] was lower than expected, while the rate of inflation and the unemployment rate were higher. The momentum of rising costs and prices, a legacy of the long inflation, proved to be extremely powerful." These sentiments were echoed by the congressional Joint Economic Committee, which had warned in 1970 of the prospect of higher unemployment, a stagnant economy, and inflation that would continue unabated. "Events during 1970 had justified these fears," it said.

THE YEAR 1970 would end on a sour political note for the president. He had hoped that Republicans might recapture one of the two houses of Congress, but the midterm elections returned Democratic majorities to both chambers. Not surprisingly, Nixon and his team concluded that the poor economy was responsible. Already nervous about the presidential election of two years hence, Nixon knew that some big changes needed to be made to his economic game plan.

By the end of his second year, Nixon abandoned gradualism altogether. In the trade-off between suppressing inflation and creating more jobs, he chose jobs. He could not stomach the political implications of unemployment. It had long been an article of faith for him that in politics, jobs counted for more than anything else. On Wednesday, November 18, 1970, for example, Chief of Staff H. R. Haldeman wrote in his diary that Nixon told him they could not afford an economic downturn, with the consequences of job losses, no matter how much inflation flared up.

In his *Economic Report to Congress* in February 1971, Nixon said, "We are facing the greatest economic test of the postwar era.

It is a test of our ability to root out inflation without consigning our free economy to the stagnation of unemployment." In early 1971, *Newsweek* wrote, "If there is not a sustained pickup in the months ahead, the economy could turn out to be Nixon's Vietnam."

Nixon wrote in his memoirs, "The first months of 1971 were the lowest point of my first term as president. The problems we confronted were so overwhelming and so apparently impervious to anything we could do to change them that it seemed possible I might not even be nominated for reelection in 1972." Indeed, the consensus among economists and Wall Street traders and investors was that in order to win in 1972, Nixon would push for anything—more spending, lower taxes, easier money—in order to generate an economic expansion. But this was exactly what foreign holders of dollars were afraid of, because looser interest rates and bigger budget deficits meant more inflation, bigger balance-of-payments deficits, and more dollars moving out of the United States to where investment opportunities were better. In short, it meant an oversupply of dollars abroad, all of them losing value.

A Run on the Dollar?

Because the United States was obligated to exchange dollars into gold at $35 an ounce when asked to do so by foreign governments and central banks, a big question became this: under what set of circumstances would they feel compelled to make such a request? There were many possibilities. Foreigners might ask for gold if they thought too many dollars were in circulation—in other words, if there existed an oversupply that made each dollar less valuable as time went by. They might want to exchange their dollars for gold if they saw that the purchasing power of their dollars was being diminished by inflation. Or they might want to do it if they thought the United States would devalue the dollar against gold, therefore rendering their dollars convertible into less gold tomorrow than today.

Whatever the circumstances, a potential crisis was imminent because the United States simply didn't have enough gold to make the exchange at $35 an ounce. Were there to be this kind of run on the bank, the United States would have to renege on a major international obligation it had pledged to uphold, with unknowable consequences for the stability of global markets and for all its broader treaty commitments.

FOREIGN GOVERNMENTS AND central banks had no desire to bring down the dollar, which, after all, was at the heart of global finance and which had been at the center of their spectacular recovery from World War II. The central banks, which held the bulk of all reserves, were a particularly tight-knit group that had worked closely together to deal with a number of currency crises in the 1960s. They were committed to a stable dollar and the preservation of the Bretton Woods system. Nevertheless, they were in a bind. Bretton Woods not only mandated that the dollar be fixed to gold, but it also required that every major country's currency be fixed to the dollar. That didn't happen automatically. The fixed rates had to be maintained by the buying and selling of currencies such that the supply and demand resulted in a certain equilibrium that equated to the desired exchange rate relationships. This buying and selling was not done by governments alone. Private traders and investors participated, too.

As the 1960s progressed, the size of the private capital markets grew dramatically, along with the expansion of the international economy. This meant that private capital became a major factor—perhaps *the* major factor—in roiling the stability of currencies. In Europe, the formation of massive new pools of dollars—called "Eurodollars"—was particularly active. These were dollar funds held abroad in foreign banks and not regulated either by the United States or foreign governments. Eurodollars could and would be used to buy and sell currencies, allowing speculators to create havoc for the efforts of the United States and Western European countries trying to keep their currencies linked to one another at the fixed Bretton Woods rates. In effect, central banks were in a battle with private speculators to keep currencies on an even keel.

For example, if private traders and investors believed too many British pounds were in circulation relative to the amount other governments and businesses wanted to hold, they might sell pounds, putting downward pressure on the currency. This would force the British government to redouble its efforts to push up the pound to Bretton Woods levels. London would have a number of options, all extremely costly. It could use its limited dollar reserves to buy pounds and lift their value. In opposing speculators, many with deep pockets, it could deplete its reserves and even borrow dollars from other governments to keep buying pounds until reaching its goal of aligning the pound with the dollar. Alternatively, the British could prop up their currency with the use of higher interest rates to make the pound more attractive to investors, thereby raising its value. But that would slow down the growth of the economy and throw millions of men and women out of work. Or Great Britain could recognize reality and agree to formally devalue the pound. However, cheaper pounds would make the United Kingdom's imports more expensive, hurt British consumers, increase inflation, and pressure labor unions to push for higher wages and benefits. Moreover, devaluation would be a political humiliation, signifying that the government had lost control over its policies and succumbed to the dictates of international financiers.

Here is another example of how the world of currency management worked. Suppose international markets thought that German marks didn't reflect the strength of the West German economy and that it was inevitable that more investors would want to hold more marks than they currently did. Speculators would then be convinced that the value of the mark would rise. They would buy marks to take advantage of their rising value. The more money that flowed into the German central bank to buy marks, the more demand there would be for marks and the stronger the currency would become. An upward spiral would ensue. West Germany would have few good options. It could use its marks to buy dollars, thereby weakening its own currency and strengthening the greenback. That could be an expensive proposition, especially if there were a strong demand for marks and weak sentiment regarding the dollar. Alternatively, it could take the temporary step of allowing its currency to float against the dollar, see how high the market pushed it, and eventually fix a new level—in other words, revalue the currency. A revaluation would not be welcome news for West Germany, either—at least not to most Germans. While a stronger mark would mean that imports would be cheaper, thus helping to hold down inflation, it would also mean that domestic industries and German workers would face new competition. And exports—on which West Germany was so dependent for its prosperity—would become more expensive to sell.

Bretton Woods allowed for devaluations and revaluations, but only in extraordinary situations and only within rules and processes governed by the IMF. The more such fundamental shifts in currency values occurred, the more doubt was cast on the viability of the fixed exchange rates established at Bretton Woods. And because the dollar was the center of the system, any big global financial disturbance would raise questions about the value of the dollar itself.

WHEN THE GLOBAL supply of dollars expanded too fast, as was the case throughout the 1960s, the central banks of Japan,

West Germany, or of other countries had to buy greenbacks to lift the dollar's value so as to maintain the link between the dollar and their currencies. In order to buy dollars, central banks often printed more of their own currency—more yen, marks, liras, pounds—to make the purchase. For Tokyo, Bonn, Rome, London, and the others, this action could be inflationary, forcing these governments to lift interest rates or tighten their budgets to try to slow growth, threatening jobs in the process. Thus, when too many dollars flowed abroad, foreign countries were put in an excruciating position. The stakes were not just about their currency values, but about their economic growth prospects, their rate of inflation, and their ability to maximize their employment.

No wonder finance ministers and central bankers abroad were angry at the United States' failure to better manage its inflation and its balance of payments.

AS 1971 BEGAN, America's rising budget deficits and low interest rates combined to produce larger balance-of-payments deficits, resulting in more dollars being sent abroad than coming in. Part of the problem was an expansive U.S. budget, which increased federal spending while not raising taxes. The result was to stimulate domestic demand and pull in more imports, thus weakening the trade balance. At the same time, inflation undermined the competitiveness of U.S. exports, slowing sales overseas. On top of all that, low interest rates in the United States caused capital to flow out of the country to places where it could earn higher returns.

To understand what happened, think of the U.S. balance of payments as consisting of two major accounts: the current account and the capital account.* Traditionally, the United States had run a big surplus in its merchandise trade—electrical products, farm machinery, food, minerals, etc. That surplus represented the United States' competitive and technological edge, with all its resources and products in demand around the world. Throughout the 1960s, the merchandise surplus was so large that it could offset all the money that flowed out of the country due to foreign aid and expenses for troops stationed abroad. In other words, the U.S. trade performance—its excess of exports over imports—more than financed its foreign aid and military commitments. The real balance-of-payments problem had been in the capital account, where billions of dollars were being sent abroad by multinational companies, long-term investors, and short-term speculators. This capital outflow—from the speculators, in particular—was sensitive to relative interest rates between the United States and its trading partners. When rates were lower in the United States than abroad, money flowed out. When they were higher in the United States than overseas, the flow was reversed. U.S. officials always felt that with the right fiscal and monetary policies, the outward short-term flows could eventually be reversed. And in fact, years went by when lots of money flowed out of the United States and then, in the following years, lots came back in.

The alarms started to ring in 1969. That year, the traditional merchandise trade surplus began to shrink. The United States had run a surplus in its merchandise trade since 1893, and the new pattern amounted to a major negative structural change in

---

* The broadest measure of trade balance is called the "current account." It includes trade in goods and services plus some other items such as net earnings on foreign investments and transfer of other funds such as foreign aid. I have chosen the term "merchandise trade balance" instead of "current account balance" for simplicity.

America's trading position. Fears arose at home and abroad that the United States was losing its basic competitiveness. This meant that short of fundamental reforms that seemed nowhere on the horizon, the United States would be bleeding dollars into the world economy with no end in sight. The challenge was acknowledged as Nixon took office. On Saturday, January 18, 1969, a transition task force chaired by Gottfried Haberler, a Harvard professor, issued this warning about the merchandise trade deficit: "Such a deterioration might well trigger an acute confidence crisis and a run on the dollar," he warned.

Logically, foreign governments, faced with the specter of out-of-control inflation in America, and confronted with endless U.S. balance-of-payments deficits, would have rushed to exchange their unwanted dollars for gold. They hesitated for a number of reasons. For one thing, they didn't want to antagonize the leader of the free world, the country that was their military defender in the Cold War. For another, Washington kept promising that it would get its economic house in order, and most financial officials abroad preferred to wait and see if the United States would make good on its latest promise, rather than to take any action. And ultimately, they knew the United States didn't have anywhere near enough gold collateral, and they had no desire to set off a frenzy that would result in a global financial panic and a political crisis in American alliances.

In the mid-1960s, financial officials thought one solution might be to create an alternative to the dollar. The idea was to allow the IMF to issue its own currency, called a "special drawing right" (SDR). Eventually, they thought, the SDR would take over from the dollar, or at least reduce the burden on the greenback. The SDR was also expected to be a potential substitute for gold. For various reasons, though, the IMF currency did not succeed as its creators had hoped. In the end, there seemed to be no substitute for the dollar. The world had become too dependent on using it for trade, investment, financial reserves, and pure accounting purposes.

IN 1970 AND early '71 the international monetary situation continued to deteriorate. Low interest rates in the United States relative to other countries drove large amounts of short-term capital abroad. An outflow of $10 billion in 1970 had increased to $30 billion in 1971, forcing U.S. allies to absorb ever more dollars and raising the most worrisome questions to date about the viability of the dollar–gold link. "I held to the hope that a major crisis could be avoided, but that hope dwindled during the winter and spring of 1971," wrote Paul Volcker, looking back years later on this period.

WITH FEARS OF an international run on U.S. gold, a wide swath of officials in Washington began to wonder whether the Bretton Woods system for currencies was broken and whether the whole setup for currencies and trade needed to be rethought. This led to broader questioning of some of the United States' most fundamental beliefs about the postwar order, including the desirability of unfettered expansion of trade and capital flows and even whether the costs of stationing U.S. troops abroad was simply too much for Washington to bear. The protectionist trade bills kept coming, one more restrictive than the next. Although it opposed most of these protectionist measures, the Nixon administration was determined to strong-arm its allies to open up their economies wider and farther than ever so that the U.S. merchandise trade deficits could be eliminated. "Western Europe and Japan were now considered not so much trading partners but rivals to be subdued," British journalist Henry Brandon wrote. "The anxieties about the huge balance of payments deficit, the enormous dollar balances that have accumulated in foreign countries, the stubborn unemployment at home, all have changed the confident attitude of Americans toward their economic future and well-being."

IN EARLY MAY 1971, the long-feared crisis erupted. It wasn't the first time—there had been a few crises in the 1960s—but during those earlier periods the underlying foundation of the monetary system had not been so weak, and the determination of Washington to pull out all stops to avert a crisis was much stronger.

A big problem emerged in West Germany and involved everything that has just been discussed—loss of foreign confidence in the dollar resulting from inflation and drain on the merchandise trade balance, the pressure that private capital exerted on the value of currencies, and the dilemmas that foreign governments and central banks faced in defending their currencies in light of the Bretton Woods fixed-exchange-rate system.

On Monday, May 3, 1971, a group of prominent West German economists called for a revaluation of the West German mark. That action was a rare occurrence but was permitted under Bretton Woods. And because West Germany had such a large trade surplus, it was actually doing the right thing by revaluing, because a stronger West German mark would likely slow the country's exports, increase its imports, and get its trade in better balance. Increasing its imports would, of course, help other countries that were running trade deficits by expanding their exports. That's how the system was supposed to work; when one part was out of whack—in this case, West Germany—there had to be some sort of adjustment in the interest of the entire system. But in the spring of 1971, the West German revaluation, and particularly the way it happened, also augured a major disruption in the Bretton Woods regime of fixed exchange rates.

On Tuesday, May 4, huge numbers of dollars poured out of the United States and the Eurodollar markets—$1.2 billion in private money, to be exact—and flowed into the Bundesbank, West Germany's central bank. It was a spectacular amount of capital for the time and a signal that markets had more confidence in the West German currency than in the U.S. dollar. It was a bet, too, that the West German currency would become even stronger in the future, that the dollar would be worth less, and that now was the time to buy marks and take advantage of the currency's likely increased value in the days and months ahead. As the mark gained strength, the dollar became weaker.

The next day, the Bundesbank took in $1 billion in the first *forty minutes* of the trading day. This was a panic, and Bonn was forced to shut down its foreign exchange markets. More dollars then flowed out of the United States and out of the reserves of other governments and into other countries, such as Austria and the Netherlands. This was again a sign of diminishing confidence in the dollar. In short order, West Germany decided to float the mark to see how high the market would push it. Pressure was building in Japan to revalue, setting off additional flows of speculative funds going to Tokyo. Several countries simply refused to buy dollars at all. All this was a far cry from the stable monetary affairs that Bretton Woods had envisioned, let alone the stabilizing role the dollar was supposed to play in the international financial system.

Meanwhile, throughout the first two weeks of May, the Netherlands, then France, and then Belgium demanded to exchange their dollars for gold. This was a logical move given that the dollar was weakening. The amounts were small, but with dwindling U.S. gold reserves, the prospects of more such transactions created additional anxieties of a gold rush and a subsequent forced closing of the U.S. gold window.

On May 12, New York senator Jacob Javits, a centrist Republican and key voice on the congressional Joint Economic Committee, proclaimed on the Senate floor that the U.S. pledge to exchange dollars presented by central banks for gold was obsolete. He suggested that Treasury stop sales of gold altogether, and he called for a world monetary conference to create a new monetary system in which the dollar would not be tied to gold. Many foreign officials, as well as traders and investors around the world, wondered if Javits was sending up a trial balloon for the Nixon administration. Although the U.S. Treasury vehemently denied that this was the case, the fact was that foreign officials and the world's investors and traders looked to the United States to assume leadership in a crisis. This could have taken the form of calling a meeting of finance ministers and central bank governors to decide on a coordinated means to quell concerns over currency gyrations. At a minimum, the U.S. treasury secretary could have made a statement that Washington was confident that any problems could and would be handled. But Washington said nothing, and a general fear arose that this benign neglect could be a precursor to America's abandoning the dollar–gold link and, with it, the stable linchpin of the entire monetary system.

On May 11, West German chancellor Willy Brandt wrote to Nixon saying that the currency crisis was causing his country great difficulties. On Monday, May 17, Paul McCracken, chairman of the Council of Economic Advisers, wrote Nixon that "There is a widespread concern in Europe that we have no interest or concern about the international financial and economic system.

Some assume we are wholly absorbed in domestic matters," he said. In the middle of this crisis, Charles Coombs, head of foreign trading at the Federal Reserve Bank of New York, wrote to his colleagues, "Traders all over the world sensed a total breakdown of policy coordination between the United States and its trading partners." Indeed, it seemed like the multilateral financial system was falling apart and that a collapse of that dollar-centric order was imminent.

THROUGHOUT LATE MAY, June, July, and the first half of August, a whirlwind of meetings within the Nixon administration would take place to avert the looming financial catastrophe. The pace of activity would accelerate in the second week of August under severe financial and political pressures. The ultimate decisions about what to do would be made during the weekend of August 13–15, 1971, at Camp David.

II.

THE CAST

4

Richard M. Nixon

As they approached the moment of truth concerning the fate of the dollar and the international monetary system in the spring of 1971, Richard Nixon led a team of men with diverse backgrounds and economic philosophies, each with strong views about America's role in the world generally and about the dollar in particular. It fell to the president to forge a consensus that overcame several underlying tensions within his team.

Some of these men, such as Treasury Secretary John Connally, were unapologetic nationalists, while others, such as National Security Advisor Henry Kissinger, wanted to forge stronger ties with America's allies. Some, like Treasury Undersecretary Paul Volcker, wished to substantially preserve the kind of fixed-exchange-rate regime embodied in Bretton Woods, while others, like Office of Management and Budget (OMB) director George Shultz, were set on substituting a free-market, floating-exchange-rate system.

Not only did their ideologies conflict at times, but so, too, did their priorities. The assistant to the president for international economic affairs, Peter Peterson, was focused on America's future competitiveness, while John Connally was riveted almost exclusively on solving the problem at hand. And then there was Federal Reserve Chairman Arthur Burns, whose positions ranged far and wide, embracing fixed exchange rates, supporting wage and price controls, and espousing deep multilateral cooperation with other central banks. It was because the Fed was an independent institution and not part of the administration that Burns occupied a singularly unique and important position in the government. Independence meant that the Fed reported to Congress and not to the administration and that, by custom, neither branch of government would interfere with the central bank's policies concerning the setting of interest rates or controlling the supply of money. Also, independence meant that the Fed chairman and other board members could speak their minds publicly without government clearance. Burns's position, plus his exalted professional reputation in domestic and global markets, made his support essential for whatever Nixon chose to do.

Not counting Burns, who in 1971 was sixty-six years old, it was a young group, average age forty-six. All but Connally were technocrats—talented, experienced, and otherwise well qualified for their jobs. Four had PhDs (Shultz, McCracken, Burns, and Kissinger), two had master's degrees (Peterson in business; Volcker had completed everything but a PhD dissertation in political economy), and two had degrees in law (Nixon, Connally).

Except for Nixon and Connally, none of them had ever been elected to political office, none had the charisma or the ambition to be a hand-shaking, media-savvy politician. When it came to judging what the domestic political traffic would bear at home, Nixon had his own strong views, of course, but otherwise relied solely on Connally.

For all, it was an article of faith that they were helping to shape and lead the "American Century," the term often used to describe an era in which the United States would dominate the world militarily, politically, economically, and culturally. Three of Nixon's top advisors—Connally, Shultz, and Kissinger—had been in World War II. All of them had been influenced by the war's aftermath, during which America led the free world to recovery from devastation and extraordinary prosperity and established and strongly supported several major global institutions such as the United Nations, the World Bank, and the IMF. Every member of Nixon's team had strong ambitions, and each had a long professional future ahead of them.

Another point about Nixon: Although any president is imbued with incalculable power in the foreign policy and national security arenas, when it came to the dollar and international finance, the president needed a much broader base of support to act effectively. After all, America's chief executive was dealing with a world of independent central banks, countless finance ministries, and millions of traders, bankers, and investors, all moving to the rhythms of the market and not just to political commands. In attempting to change not just the relative value of the dollar but the underlying system on which its value was based, a president had to be able to act with confidence and credibility. So, no matter how powerful Nixon was, no matter how strong his leadership could have been, in the end he needed a relatively unified team behind him.

Key members of Nixon's entourage were sworn in with the new president on Tuesday, January 21, 1969, the day after his inauguration. It was not a radically conservative group. "It is now evident that the President-elect has selected a Cabinet and a White House staff which also reflect the political center," wrote Roscoe Drummond, a *Washington Post* columnist. "This is where the votes are, present and future. And this is where Nixon himself comfortably stands."

UNDERSTANDING THE RICHARD Nixon of this August weekend requires not yet equating the president with the abuses of power and the obstructions of justice—including the lies, the break-ins, the vendettas, and the cover-ups—that defined his administration as it abruptly ended in 1974, when he became the first and only president ever to resign from office. Let us then focus on Nixon in 1969, 1970, and 1971—Nixon "before the fall," to use the title of a book about him before the Watergate scandal by William Safire, one of his speechwriters.

Nixon's rise from humble beginnings had been swift. He came from a Quaker family in the small town of Yorba Linda, California, working his way to a law school scholarship at Duke University, where he graduated third in his 1937 class. During World War II, he became an operations officer in the South Pacific Combat Air Transport Command overseeing logistical issues but not being involved in combat.

After the war, Nixon ran for U.S. Congress against a five-term incumbent, displaying the ambition and aggressiveness that would mark his political career. His campaign leaned heavily on Communist fearmongering, the most charged political issue of the day. He also campaigned on a platform of serving the "forgotten man," what he described as the quiet people who work hard, struggle to support their families, and don't make big demands on government. Allegiance to this group and denigration of the elites and the eastern "establishment" would become one of the consistent themes of his political career. Those whom Nixon particularly abhorred ranged from Ivy League graduates, to the eastern media, to the upper echelons of the State Department and CIA, to big business and Wall Street executives, to those who were regulars on the social scene in the fashionable Georgetown district in Washington.

Nixon quickly made a name for himself in Congress as a dogged legislator and ruthless competitor. After two terms, he won a Senate seat, and in 1952, the thirty-nine-year-old senator was offered the vice-presidential spot on the Dwight D. Eisenhower ticket. Eisenhower, trying to stay above the political fray as befitted a World War II military legend, needed a street fighter on his team, and Nixon filled the requirement. As vice president from 1953 to 1960, Nixon's most active role was in foreign affairs. He participated in deliberations of the National Security Council, visited over fifty countries, and built up personal associations around the world and throughout the United States. It would be in the international realm that he would focus so much of his ambition and attention.

Nixon won the Republican nomination to succeed Eisenhower as president in 1960 but narrowly lost the election to John F. Kennedy. By today's standards, the differences between the two candidates' political positions were not great. When it came to the relatively new medium of television, however, Nixon was no match for the stylish and charming Kennedy. Another reason Nixon may have lost the election, and one he would always remember, was that the economy slowed down in the last months of the campaign. Indeed, a mild recession took place. Despite Nixon's pressure on the Federal Reserve, Chairman William McChesney Martin refused to lower interest rates to boost Nixon's prospects. From then on, Nixon was suspicious of the merits of an independent central bank, and he never forgot that unemployment and recessions, far more than other issues such as inflation, were the bane of incumbent politicians running for reelection.

In 1962 Nixon, still licking his wounds, ran for governor of California against incumbent Pat Brown and lost. Everyone considered his political career over, and he moved to New York City to join a prominent law firm. But when, in 1964, Republican candidate Barry Goldwater lost in a landslide to Lyndon Johnson, Nixon saw the possibility of a comeback. He campaigned for Republican candidates in the 1966 midterm elections and began laying the groundwork for another presidential run. In the summer of 1968, when Nixon won the Republican nomination for president, James Reston of the *New York Times* wrote, "It is the greatest comeback since Lazarus, and even in this mean and vicious business there is scarcely a Nixon doubter who does not recognize a remarkable political achievement." That November, he won the presidency.

NIXON CONSIDERED HIMSELF a strategic thinker, an accurate self-perception. As president, his goal was no less than to replace the sterile and dangerous bipolar Cold War between the United States and the USSR by bringing China into the global power mix. Nixon saw, as many at the time did not, that the two Communist powers were hardly allies, but were instead deeply wary of one another. Preying on those anxieties, Nixon launched separate diplomatic overtures to both, playing the two powers off each other in what he called "triangular diplomacy."

Beyond reshaping big-power relationships, a major part of Nixon's preoccupation was lessening the burden on the United States to lead and manage so much of the free world's affairs. Within his first seven months in office, he announced the Nixon Doctrine, which had implications for every aspect of American foreign policy, trade, and policies toward the dollar.

The new policy was unofficially launched in late July 1969, in the midst of an exhausting ten-day whirlwind trip that would take Nixon to Guam, Thailand, Vietnam, India, Pakistan, the Philippines, and Indonesia. On Friday, July 25, he talked to reporters at the Guam Officers Club on a not-for-attribution basis. One of them asked him if the United States would come to the aid of another Asian country aside from Vietnam. Nixon gave a long answer that implied that unless a country were attacked from a foreign power, or the United States had a defense treaty with the country, Washington would not be providing soldiers but only financial and technical support.

At first, the new Nixon Doctrine seemed to pertain just to Asia. But in a few weeks the administration touted it as a change of course for U.S. foreign policy generally. The new focus was a direct contrast to the policies of Presidents Kennedy and Johnson, who acted as if there were no limits to America's commitments to peace, stability, and prosperity around the world. It would amount to a scaling back of what the United States was willing to do for other countries. For some observers, it even amounted to an American retreat from world responsibilities. Years later, Henry Kissinger recalled that the Nixon Doctrine became a coherent statement to explain that the United States understood the danger of overextending its resources.

Aside from its direct relevance to fighting wars, the administration began to focus on three other realms of policy where it would aim to spread the costs and responsibilities of managing the free world more evenly among the allies.

To start with, Nixon underlined the need for Western Europe and Japan to spend more for their national defense, including paying for the stationing of U.S. troops on their soil and for a larger share of the cost of American military equipment used to protect them. Washington also wanted the allies to procure more of their defense needs from U.S. suppliers. In all cases, the purpose was to alleviate the pressures on the United States' budget and its merchandise trade balance.

In the arena of trade, Nixon was saying that America would no longer tolerate the situation of its markets being more open to its trading partners than vice versa. Such uneven bargains may have been acceptable in the 1950s and early 1960s, the administration said, but that was at a time when Western Europe and Japan desperately needed access to U.S. markets for their recovery from the war. Also, that was a time when the United States was so competitive and prosperous that the unbalanced nature of trade didn't really matter. Those days were over.

The Nixon Doctrine also had strong implications for the dollar and global monetary affairs, above all for the U.S. commitment to exchange dollars held abroad for gold at $35 an ounce, which had become a debilitating burden. Nixon officials believed that the dollar was significantly overvalued at that exchange rate and that the currencies of West Germany, Japan, and some other countries were undervalued. They wanted to rectify this imbalance by forcing a onetime change in the relationships among the currencies of these key countries. Put simply, Washington wanted the dollar to be weaker relative to the mark and the yen.

Why was overvaluation of the dollar such a problem? Because it inflated the price of U.S. exports—say, grain or machinery—for European and Japanese markets, thus slowing U.S. sales abroad. And it made the prices of foreign goods—say, shoes or televisions from Japan or steel and autos from West Germany—cheaper for Americans, thus accelerating the flood of imports into the United States and worsening the trade balance. It is true that foreigners already had an advantage because of their lower wage rates coupled with their rising levels of productivity, but for the

United States, the super-strong dollar amplified this competitive problem.

Because the overvalued dollar slowed exports and increased imports, and because more capital was flowing out of rather than into the United States, the balance of payments widened. In theory, there were several ways to deal with this problem, all extremely difficult.

In the first instance, the United States could have demanded that its major trading partners, consisting primarily of Japan and Western Europe, open their markets to U.S. exports in orders of magnitude and at speeds well beyond what these countries were considering. It would have taken massive political pressure on the allies to get them to move, and maybe even that wouldn't have worked, given that trade was so important to these countries' economies. Failing that, the United States could have decided to block its trading partners' exports from entering its market. But that would have meant going protectionist, a move that was counter to Nixon's philosophy.

If dramatic trade remedies were not politically feasible, the United States could have considered devaluing the dollar against gold. Instead of an ounce of gold being worth $35, the new price could be, say, $38 per ounce, making every dollar worth less in terms of gold. The impact here would be to make U.S. exports cheaper in international markets relative to the price of foreign products, and also to make imports more expensive, thus improving the U.S. trade balance. The problem with this solution, however, was that the dollar was the centerpiece of the global monetary system. It was the sun around which planets revolved. There was no provision of Bretton Woods that contemplated that the central currency could be devalued, and doing so was a dangerous step into the unknown that could have led to financial chaos. Besides, there was another problem. Every other currency was fixed to the dollar at a set price. Washington was afraid that if it devalued the dollar, every other currency would be adjusted downward, too. That would mean that the United States would gain no advantage, while the entire Bretton Woods system could be destabilized. If the United States wanted a cheaper dollar, the only option was to use the full force of its diplomatic and military leverage to get its partners not to devalue alongside the dollar, and that risked blowing up critical political alliances in the middle of the Cold War.

Rectifying the balance of payments without major changes in the trade and currency arenas meant the United States would have to raise interest rates to keep dollars from going abroad and to attract other funds into the United States. Also, it would have to slash its budget to slow down growth and reduce demand for imports. Taken together, elevating interest rates and constricting fiscal policy might help the balance of payments, but they would also push up unemployment. Nixon would not tolerate that.

The bottom line was that America faced the need for a broad-based adjustment of burdens among the allies. This would involve sharing military and economic burdens in ways never before contemplated. "The [Nixon] Doctrine set a tone and carried an implicit warning," wrote British journalist Henry Brandon. "It was a very clever way of telling the world that an end of an era had arrived."

IN FOREIGN POLICY, both custom and the Constitution gave Nixon exceptional power to act alone, to be decisive and to make dramatic decisions without first getting congressional approval. Not so in the economic sphere, where authority was widely shared with Capitol Hill, and often with state governors, and where advanced consultation with business and labor leaders could be essential, too. What's more, economic issues didn't excite Nixon. He told journalist Theodore White in 1967, "I've always thought the country could run itself domestically without a president . . . [but] you need a president for foreign policy." Herbert Stein, another of Nixon's top economic advisors, said that Nixon "felt that for Republicans, economics was not a winning issue because their conservatism was at odds with popular expectations of what government should do." Also, Nixon was always looking for a major dramatic breakthrough, which was something almost impossible to do in domestic economic policy, where so many constituencies had to be brought on board in advance.

Over the next three years, Nixon would be obsessed with avoiding recession and the higher unemployment it would bring. He was convinced that jobs equaled votes and that no one ever lost an election because of rising prices. Biographer Richard Reeves wrote that "Nixon repeatedly interrupted Cabinet meetings to go over the history of Republican defeats when the economy was in slow growth or decline." "When you start talking about inflation in the abstract, it is hard for people to understand," Nixon told a cabinet committee the month after he took office. "But when unemployment goes up one half of one percent, that's dynamite."

Another of his strongest convictions was opposition to mandatory wage and price controls, which he never tired of pooh-poohing. In his January 27, 1969, press conference he said, "I do not go along with the suggestion that inflation can be effectively controlled by exhorting labor and management and industry to follow certain guidelines . . . for the primary responsibility for controlling inflation rests with the national administration and its handling of fiscal and monetary affairs." Later in 1969, he explained his aversion to such controls. "My first job in government was with the old Office of Price Administration at the beginning of World War II," he recalled in an address to the nation on

Friday, October 17, 1969. "And from personal experience, let me just say this: Wage and price controls are bad for business, bad for the working man, and bad for consumers. Rationing, black markets, regimentation—that is the wrong road for America, and I will not take the nation down that road."

HIS DISTASTE FOR economics extended to the international realm, too. Other than the bromides of keeping a strong dollar, supporting free trade, and encouraging the free flow of capital, he said almost nothing about international economics during the presidential campaign. Before the election, economist Paul Samuelson wrote in *Newsweek*, "A question I am increasingly asked is, 'Will President Nixon devalue [the dollar] and throw the blame on his predecessors?'" Samuelson answered his own question by saying only someone with enormous confidence would take such a risk and that Nixon had given no indication he was such a person. Shortly after the election, Samuelson wrote, "President-elect Nixon is going to find himself doing pretty much what his two predecessors were doing. This is guaranteed by the closeness of the election." Samuelson would prove wrong on both counts.

Nixon's main concern was to do whatever was necessary to keep economic issues from undercutting his real passions: arms control negotiations with the Soviets, future openings to China, and other big, traditional foreign policy initiatives. On Monday, March 2, 1970, for example, he wrote a memo to chief of staff H. R. Haldeman: "I do not want to be bothered with international monetary matters," he said. "Problems should be farmed out . . ."

Indeed, his relative lack of interest in the dollar could be seen on the eve of his taking office, when, unlike for previous presidents, the dollar seemed low on his agenda. Nixon had appointed Chicago banker David Kennedy as his first secretary of the treasury.

At a press conference in December 1968, over which Kennedy presided, a reporter observed that LBJ had been quick to give a strong commitment to preserve the dollar–gold link. What's Nixon's position on gold? the reporter asked. Kennedy responded, "I would like to defer because we are going to have a discussion on the whole international situation later. I want to keep every option open." This statement appeared at first to be a major change in the United States' twenty-five-year policy of unambiguously supporting the dollar–gold link. Herbert Klein, Nixon's designated communications director, jumped in to rescue Kennedy, telling reporters that Kennedy was not talking about a change in policy. By the next day, markets were still wondering whether something big had changed, and Ron Ziegler, Nixon's press secretary, tried again to walk back what Kennedy had said. "We do not anticipate any changes in the price of gold," he said (which was another way of saying that the dollar would not be devalued). But the damage to foreign perceptions of Nixon's dedication to the dollar–gold link had been done.

To the extent he thought much about it, Nixon believed that gold was an anachronism in monetary affairs and that the entire international monetary setup needed to be rebuilt to avoid recurrent crises. Speaking at the Treasury Department a few weeks after the inauguration, he said, "Now is the time to examine our international monetary system to see where its strengths are, where its weaknesses are, and then to supply the leadership which is responsible, not dictatorial, leadership that looks to the good judgment and good advice that we can get from our friends abroad." In private, he disparaged Treasury officials, who he felt were too wedded to Bretton Woods and resistant to any new ideas.

When it came to international commerce, Nixon talked often about free trade, but he made it clear that there had to be some exceptions, such as protection for textiles. "I believe that the whole world will be served by moving toward freer trade than toward protectionism," he said in his first high-profile statement as president on the subject, on Thursday, February 6, 1969. "As far as the textile situation is concerned, that is a special problem which has caused great stress in certain parts of the country."

Nixon was often critical about the behavior of other countries in private and willing to think about wielding a big political club against the United States' economic rivals. Three months after taking office, for example, he told the cabinet that his support for NATO would be jeopardized if Western Europeans blocked U.S. exports of soybeans. He said that if the European Community turned inward, it could forget U.S. support.

NIXON WAS LESS of an ideologue than his opponents made him out to be. He often appeared to be a conservative with liberal views. During his time in office, social programs increased as a percentage of overall government spending from 33 percent to 50 percent. Speechwriter William Safire said that Nixon was often labeled "unprincipled." But this charge was made by the right, who felt he wasn't conservative enough, and by the left, who resented that he had co-opted some of their ideas, ranging from protecting the environment to expanding federal support for the arts. Biographer Richard Reeves tells of how Arthur Burns, two months after the 1968 election and then counselor to the president, was arguing with Daniel Patrick Moynihan, a prominent domestic policy advisor to Nixon, about welfare reform plans. Burns was saying that Moynihan's liberal ideas were not consistent with Nixon's philosophy. John Ehrlichman, chief domestic policy advisor to Nixon, was listening to the two men arguing.

He had had enough of this debate. "Don't you realize the President doesn't have a philosophy?" he asked.

Nixon wanted to mesmerize the world, especially his political opponents, with big, bold surprise initiatives. "I have made some bad decisions," he once said, "but a good one was this: When you bite the bullet, bite it hard—go for the big play." Henry Kissinger once said that Nixon had a motto that went, "You pay the same price for doing something halfway as for doing it completely. So if you do something, you might as well do it completely." Kissinger explained that this meant that once Nixon was convinced of a course, he would try to take the most sweeping solution presented to him or that he could otherwise invent.

Nixon was a loner. He felt ill at ease with almost everyone but his wife and daughters. Safire describes his awkwardness around even his closest associates. It was January 20, 1970, the anniversary of Nixon's first year in office. You could have imagined some sort of congratulatory ceremony with his staff, but instead, late in the day, Nixon called in H. R. Haldeman, his chief of staff, and Rose Mary Woods, his longtime and trusted secretary. The lights were turned down in the office, and Nixon was wearing his overcoat, preparing to leave for the day. As Woods and Haldeman entered his office, Nixon opened a little music box. Out came the tune of "Hail to the Chief," with the music slowing at the end as though it had been played one too many times. Nixon shut the box. "It's been a year," he said, omitting any modifying adjective. And then he walked out.

In Nixon's mind, a certain aloofness was synonymous with leadership. He deeply admired the mysterious stand-alone image that French president Charles de Gaulle established for himself. Reading de Gaulle's memoirs, Nixon had underlined the sentence "Great men of action have without exception possessed in a very high degree the faculty of withdrawing into themselves." Safire wrote, "[Nixon] wanted to seem slightly out of reach without seeming in the least out of touch."

He wanted to keep his political foes off balance, too. He once told Sen. Bob Dole, "I just get up every morning to confound my enemies." Bryce Harlow, one of his closest senior advisors, put it this way: "Richard Nixon went up the walls of life with his claws."

Nixon surrounded himself with several strong advisors, most notably John Connally and Henry Kissinger. But although he seemed always in charge, his method of control was less clear. Wrote Henry Brandon regarding Kissinger in particular, "The extent to which [the president] retained command remained a mystery even to those who had the opportunity to see the two [Nixon and Kissinger] together at various formal meetings."

In addition to the overwhelming power inherent in the office, perhaps Nixon's control could also be attributed to the way he made big decisions. Virtually everything that came to him went through two assistants in the White House, H. R. Haldeman and John Ehrlichman, sometimes called by other staff "the Berlin Wall" or "the Germans." Few of Nixon's advisors other than Kissinger and Connally could walk into his office without the approval of these two gatekeepers.

"Nixon's style is that of a judge," explained *Fortune* reporter Juan Cameron in July 1970. "On important matters, he insists on carefully constructed written briefs, which often are followed with oral arguments by his staff and Cabinet officers." Then "he retires, in a manner of speaking, to his chambers . . . as he ponders his decision in solitude."

John B. Connally Jr.

In December 1968, right after his election, President Nixon chose as his secretary of the treasury David Kennedy, chairman and CEO of the Continental Illinois National Bank and Trust Company of Chicago. Kennedy had an exquisite résumé. In addition to climbing the ladder at Continental, he worked at the Federal Reserve Board as a lawyer and economist, held the job of special assistant to Secretary of the Treasury George Humphrey in the Eisenhower administration, and headed a task force on the federal budget process for LBJ. At Continental, Kennedy expanded operations in London and Japan and built extensive relationships with financial officials abroad. A moderate Republican, he was even considered for the top Treasury post in the Johnson administration. The nomination of Kennedy was consistent with the string of top financial officials at Treasury over the previous decade who were experienced, knowledgeable, and respected around the world—men such as Douglas Dillon, Robert Roosa, and Henry "Joe" Fowler. What seemed to matter most to Nixon was that Kennedy was not part of the New York–Boston banking establishment that the president so deeply resented. But, as it turned out, Kennedy and Nixon never clicked, and within a few months, Kennedy was eclipsed in terms of effectiveness by both Paul McCracken, chairman of the Council of Economic Advisers, and George Shultz, secretary of labor.

At the end of Nixon's second year, in December 1970, Kennedy was replaced by former Texas governor John Connally. Compared to the past three treasury secretaries, the new appointment was a startling contrast—no banking experience, no knowledge of international finance, few personal relationships around the world, no inclinations to find a diplomatic solution to shared concerns. Still, Connally was a perfect fit for what Nixon wanted and needed. After 1970, in fact, at virtually every critical moment, Nixon told Haldeman to check with Connally for his advice on economic matters and even on political and foreign policy issues.

ON NOVEMBER 19, 1970, shortly after Nixon's drubbing in the midterm elections, the President's Advisory Council on Executive Organization was presenting its final recommendations to President Nixon in Los Angeles. Nixon had set up this group shortly after his election in 1968 under the chairmanship of Roy Ash, cofounder and president of Litton Industries. Dubbed the "Ash Commission," it aimed to provide a road map for reorganizing the executive branch, which the new administration believed to be bloated and inefficient stemming from Lyndon Johnson's sprawling Great Society programs. The president grew bored with the briefing and what he considered to be bureaucratic minutiae rather than the grand initiatives that excited him. As Ash droned on, mired in the complexities of organizational charts, he had clearly lost his most important listener.

Suddenly, one of the commission's members rose to the rescue. John Connally, the fifty-three-year-old former governor of Texas, started to speak. The key point, he said, was that the reorganization could be seen as a grand political stroke, giving the president the aura of pushing the government into a new era of modern, more effective management and forcing his detractors to defend the status quo. It mattered less whether the initiative was supported by Congress, Connally said, for it could be a big political win just to launch it. Nixon was impressed and agreed to push the commission's recommendations with Congress.

The president did not know Connally personally, but he knew of him and admired him as an experienced, ruthless politician. Connally had been at the right hand of Lyndon Johnson from the early days of LBJ's political career—as chief aide for Johnson's election to the House and Senate in the 1940s and as the political operative behind his elevation to vice president in 1960 and to president in 1964. In 1962, Connally had been elected governor of Texas, eventually serving for three terms. He rode with JFK and was wounded in the fateful motorcade in Dallas in 1963. He also helped deliver Texas for Hubert Humphrey, who ran against Nixon in 1968, a feat Nixon had not forgotten.

The president also knew that the former governor was a conservative Southern Democrat. Connally had enthusiastically supported America's war in Vietnam. He relished challenging Democratic liberals such as Senators Eugene McCarthy and George McGovern for their opposition to it. He aggressively attacked Democratic presidential candidate Hubert Humphrey when his support for the war wavered during the 1968 campaign. And when it came to everything from civil rights to managing national social programs, Connally, in contrast to most Democrats, was an ardent believer in returning power to the states. He was a conservative who believed in active government, a mind-set not too different from that of Nixon. The Texas governor also reflected major changes in the political complexion of the nation, including the growing trend for the once-solid Democratic South to elect more Republicans. Nixon sensed that conservative Democrats like Connally were ripe for conversion to the Republican Party.

After the midterm congressional election of 1970, Nixon was unhappy with the lackluster character of his cabinet, a group that elicited little excitement from the public. He wanted to inject some new blood. "Every Cabinet should have at least one potential President in it," Nixon mused to top aide John Ehrlichman. "Mine doesn't." Nixon was already thinking about the presidential election of 1972 and the need for someone to oversee an economy that he feared could be his biggest vulnerability. The president was attracted to the idea of bringing in a star, perhaps a prominent Democrat, to shake things up.

On Friday, December 4, 1970, Nixon met with Connally at the White House and offered him the Treasury post. Connally agreed to consider it. Nixon ordered Haldeman to seal the offer. Haldeman was told to hint that there could be an even bigger position in the future—the possibility of having a shot at being Nixon's vice-presidential running mate in 1972. At breakfast with the president three days later, Connally accepted Nixon's offer.

LIKE NIXON, CONNALLY had risen from modest means, his father having been a grocer, a tenant farmer, a one-car chauffeur service, and a rancher. Connally grew up in the Great Depression and attended the University of Texas at Austin and its law school. He was a decorated naval officer—a "fighter directions officer" on an aircraft carrier in the Pacific whose job was to guide American naval pilots to intercept incoming enemy aircraft.

He had a number of qualities that endeared him to the president. A fastidious, elegant dresser, he was tall, movie-star handsome, and exuded confidence, charisma, and command. He was the personification of the state of Texas from which he came—big, brash, and flamboyant, with an outsize sense of himself. Peter Peterson, who served alongside Connally, wrote of him, "When he walked into a room he radiated ambition, self-esteem, power. Attention swung to his full and spontaneous smile, his booming voice, the big cigar, the firm handshake, the virile and commanding presence that signaled to everyone present—as he himself assumed—that the room was his." William Safire said that "Connally meant stimulation, excitement, political savvy to Nixon, and the presence of what the President liked to call a 'bold stroke.'" Henry Kissinger wrote in his memoirs, "Connally's swaggering self-assurance was Nixon's Walter Mitty image of himself."

Connally was also hyper-articulate, a master at connecting the dots and painting a big picture in simple, compelling language. At the Democratic Governors' Convention in 1966, for example, a major discussion of the Vietnam War was taking place. Grant Sawyer, governor of Nevada, was given the mandate to make the case supporting LBJ's sending more U.S. troops and escalating the fighting in Vietnam. Responding to a question, Sawyer was flubbing his way through an answer when, with no preparation, Connally intervened to save him, much as he would save Ash a few years later. Connally made a powerful case, stitching together years of American foreign policy, the risks of not drawing a line in Southeast Asia, and the dangers that Russia and China posed. "It was the most impressive presentation on Vietnam I have ever heard," recalled Larry Temple, who served as Connally's top aide before he became White House counsel to President Johnson.

All his life, Connally had amazed others with this ability to ingest massive amounts of information, synthesize it, and recall it months or years later in simple language that captured an issue's essence. Once, shortly after he became treasury secretary, Connally asked an aide to prepare a speech on trade. Connally referred to a presentation given six months earlier by the chairman of Texas Instruments, a manufacturer of semiconductors. Connally recollected to the aide a host of numbers in that speech relating to percentage changes in GNP, trade flows, workforce issues. The aide looked up the presentation and saw that, with one very small exception, Connally had recalled everything precisely.

ANOTHER ASPECT OF Connally's experience impressed Nixon, too. As chief lobbyist for Texas billionaire Sid Richardson throughout most of the 1950s, Connally immersed himself in the Texas universe of magnates who won, lost, and won again gargantuan fortunes in oil and gas. For nearly a decade, he developed strong ties to the oil and gas industry, famous for its ability to manipulate politicians at all levels. Connally was often considered the strategic and operational behind-the-scenes mastermind looking after the fortunes of Richardson and other Texas billionaires. His time with Richardson was taken up with epic fights between Texas and the federal government over controls of offshore oil fields and regulation of gas prices—issues for which vast fortunes were at stake. The association with Richardson, and the forging of his links between Texas and Washington, gave Connally an acute understanding of how to leverage money and access in dealing with some of the most contentious legislation of the mid-twentieth century.

CONNALLY WAS THE secretary of the navy in the Kennedy administration for just one year, 1961. The job gave him a first-row seat for the evolution of America's high technology. He observed that institutions such as Harvard, MIT, Caltech, and the RAND Corporation were receiving massive federal infusions of funds for research in high technology. And he saw the emergence of research clusters, such as Silicon Valley and the Harvard-MIT complex in Massachusetts.

This experience would help define his principal legacy as governor of Texas from 1962 to 1968. He wasted no time in trying to reorient the state from its roots in oil and gas, plus agriculture and a few smattered industries, to becoming a technology powerhouse. "Brains not brawn" became a tagline for his efforts. "Industry follows brainpower," he told business leaders, and he proceeded to restructure the state's entire higher education system, gathering support from business leaders and raising taxes three consecutive times in the process. This was a herculean achievement in a state historically opposed to funding social programs. Connally had proved himself not just an experienced political strategist but a skillful operative.

THERE WAS THE additional appeal for Nixon that Connally had been schooled in politics by LBJ. No one had more experience in maneuvering the seamier side of politics—including deception, blackmail, and bribery. Connally once told another political operative, "You operate like I do . . . We're termites. If the sun shines on us, we die. We do much better in the shadows." Sargent Shriver, who ran the Peace Corps under JFK and several programs under LBJ's war on poverty, once compared Lyndon Johnson and Connally. He said they were two "ambitious, political animals who came out of the same jungle . . . lions out of the same bush . . . who had the same emotions, the same aggressiveness, the same showboating, the same strengths and weaknesses."

LBJ once told George Ball, a top diplomat in the Kennedy and Johnson administrations, how he viewed Connally. "You know, George," he said, "I can use raw power as well as anyone. You've seen me do it. But the difference between John and me is he *loves* it. I *hate* it." (Ball discounted Johnson's description of himself, but he knew LBJ meant what he said about Connally.)

Connally was not reluctant to engage in the politics of personal destruction. For example, he presided over a news conference in 1960, when LBJ and Kennedy were competing in the Democratic primaries, to disclose that JFK had Addison's disease. Nor was Connally immune from major political donnybrooks. In the wake of the assassination of Martin Luther King Jr. and the rioting in response, for instance, Connally publicly disparaged the civil rights leader, saying, "Much of what Martin Luther King said and much of what he did, many of us could violently disagree with, but none of us should have wished him this kind of fate."

He was an exceptionally tough opponent and negotiator. "It's not enough for Connally to beat you," said a congressman from the state legislature who was an ally of the governor while the two held office. "He's got to rub your nose in the dirt." Early on as treasury secretary, Connally startled Henry Kissinger—no amateur infighter himself—when he told him, "You will be measured in this town [Washington] by the enemies you destroy. The bigger they are, the bigger you will be."

ON THE MORNING of Monday, December 14, 1970, the Nixon cabinet and the Republican congressional leadership were meeting in the White House Cabinet Room to hear George Shultz, the director of the Office of Management and Budget, present a new budget. Nixon made an unexpected entrance and interrupted the discussion to announce that he had decided on a successor to Treasury Secretary David Kennedy: the former governor of Texas, John Connally. He went on to describe Connally's qualifications, and then continued: "We have a Republican president. We have a Democratic Congress. The problems that we face at home and abroad . . . are not Republican problems or Democratic problems. They are American problems . . . [Because of] John Connally, we will be able to present our programs both at home and abroad, not simply as partisan programs but as programs that both Democrats and Republicans, we believe, can support."

In fact, Connally's party affiliation was complicated. He was unpopular in the Democratic Party for being so conservative and was often suspected as being a closet Republican. Yet he was seen by the Republicans as being a Trojan horse for the Democrats. Nevertheless, Nixon's admirers felt the president had pulled off a triple play. "He had repaired his relations with a hostile Congress; gained at least the appearance of bipartisan support for his economic policies, and [appointed] an articulate advocate to promote their success; and—most important—put Texas within his grasp," wrote Richard Whalen, a journalist and onetime Nixon speechwriter, in *Harper's*. Connally would also be the only senior member of the administration who was a down-and-out real, attractive politician—someone who could press the flesh when necessary, someone who could patiently listen to the complaints of people who counted, someone who could make endless small talk when essential, and someone who could consult, cajole, persuade. Nixon had no one else to do any of that.

The day after Nixon announced Connally's appointment, a White House meeting took place concerning the upcoming State of the Union address. Nixon had invited Connally to attend. It would be almost six weeks until his Senate confirmation, but Connally didn't wait to project a formidable presence. At one point, the discussion about Nixon's program for sharing federal revenues with the states became bogged down in details. Connally jumped in. He said that no one outside Washington knew what the program was, yet it held the seeds for a revolutionary and innovative change of direction in national policy. He brought up the dramatic impact of the Great Society programs on American psychology. Revenue sharing could be the core of a new future, he said, and a popular antidote to the general feeling that after LBJ, Washington now controlled too much. Connally was foreshadowing the conservative tide that was building in the country. He urged Nixon to make a much bigger deal out of turning funds back to the states. "I say let's run the risk," Connally told the group, showcasing his swing-for-the-fences mentality. "If you lose, you lose big—but what's the sense in losing small?" Nixon was delighted.

Two weeks later, Connally had told Nixon that the media should be focusing on a big political comeback story about the president. Playing to Nixon's vanity, Connally recalled that some of the great leaders of the twentieth century, such as Churchill and de Gaulle, had come back from defeat, implying that Nixon's journey was comparable and should be promoted as such. For the next few months, Connally was put in charge of a White House effort to generate more favorable stories about Nixon.

IN THE DAYS before he was confirmed, Connally demonstrated how he would operate in his new job. H. R. Haldeman had become an extremely powerful chief of staff, more powerful than most cabinet officers, controlling almost total access to Nixon from others in the administration. He called Connally to say he would send over some White House briefers to begin the treasury secretary's education in the administration's policies. Connally replied that he would call if he needed help and that if he required clarification of the administration's policies, he would get it directly from the president himself. On another occasion, Haldeman came to see Connally and was kept waiting in the outer room for thirty minutes, in a demonstration of just what Connally thought of Nixon's notorious palace guard. Connally's insistence on having direct and unfettered access to Nixon even reached into foreign affairs and the National Security Council. Usually, any memo to Nixon dealing with U.S. relations abroad went through Henry Kissinger for comment before the president read it. This made Connally furious. Often when he sent a memo, he would have his assistant call Nixon's secretary, Rose Mary Woods, to see if the correspondence had been diverted, objecting vociferously if it had been.

CONNALLY ARRIVED ON the job as a nationalist, maybe an ultranationalist, who felt that America's allies had for too long taken advantage of the country that so generously bankrolled their recovery from the war. Henry Brandon wrote that Connally couldn't care less about the positive implications of the Western European countries getting together. Instead, what he saw was a protectionist bloc out to undercut the United States. He had no attachment to the Bretton Woods structures of international finance and law; indeed, he was more apt to focus on the constraints that these institutions and principles put on the United States.

He had a remarkable lack of ideological predisposition. Everything was circumstantial. Free and open trade? A strong or weak dollar? Wage and price controls? It all depended on the politics of the moment. *BusinessWeek* once described one of his favorite parables, the story of a new preacher being interviewed by the deacons of a church for a job. One of the deacons asks the preacher, "Do you believe in the divine creation or do you hold that man descended from monkeys?" Not knowing how to answer, the preacher finally swallows hard and says solemnly, "Deacon, I can preach it either way." Connally himself was known to say, "I can play it round or I can play it flat. Just tell me how to play it."

That said, when Connally entered Treasury, he quickly developed a politician's view of the dollar. To him, the dollar was all about trade, and trade was all about jobs. Forget about the intricacies of international monetary affairs—how currencies should be valued, what the obligations of the Bretton Woods Agreement were, what role the International Monetary Fund should play, all the issues that central bank governors and most finance ministries fretted about in constant meetings. For Connally, the litmus test of the dollar was whether its value allowed the United States to compete in international markets, whether it led to a positive or negative merchandise trade balance, whether it added or took away jobs. In looking at exchange rates this way, Connally was aligned with Nixon, Congress, and the American public.

SINCE THE MID-1960s, trade had become high politics. In the U.S. Constitution, Congress was given the responsibility for international trade. In addition to holding extensive hearings, Congress had been responsible for legislation that related to several rounds of global trade negotiations. Most congressmen and senators understood tariffs and other impediments to commerce, and they were constantly approached by constituents who were either hurt by imports or who wanted a better opportunity to export. Business and labor focused significant attention on trade matters.

Currency issues were a different matter. They were the province of central bankers and finance ministers who met in different places around the world and whose activities were covered mostly by the specialized financial press. Congressional committees were not significantly focused on them; nor were major companies and unions. Currency matters seemed more technical, less political. In any event, everyone knew that the value of the dollar was fixed to gold at a set rate, so until 1971, there didn't seem much to debate.

In Connally's view, the United States was being hurt by the protectionist actions of other countries, which inflamed his nationalist instincts. After all the United States had done for other countries since the war, after all the trade concessions America had made, he found it outrageous that Western Europe and Japan still blocked U.S. exports. Connally especially resented that Bonn and Tokyo—both so dependent on the United States for military defense, no less—were amassing large export surpluses, evidence to him that they were selfish predators in the global economy and not interested in giving anything back. To Connally they were cheating, pure and simple.

More specifically, Connally resented how well Japan and West Germany were doing compared to the United States. During the 1960s, an era when manufacturing dominated economies, U.S. exports of manufactured goods increased by 110 percent, taking advantage of the rapidly growing world economy. By contrast, those of West Germany increased 200 percent and those of Japan rose by 400 percent. Japan's increasing momentum was particularly breathtaking: its exports grew by 17 percent per year in the 1960s, by 20 percent annually in the 1967–71 period, and by 24.5 percent in 1971. Taking the European Community and Japan together, merchandise imports into America had virtually doubled over the previous five years.

Connally's nationalist views found some sympathy with Nixon, even though they were a sharp departure from what had been the official trade policy of the past four administrations, all of which supported international negotiations to free up trade.

In addition to opening up new trading opportunities for American companies, trade pacts offered significant political benefits in strengthening the political ties with Western Europe and Japan. Thus, U.S. officials were reluctant to press foreign countries too hard, lest alliances be disrupted. By 1970, rising protectionist pressures in the United States reflected the fact that this delicate balance was being challenged by Congress and the labor movement, both of which said that the United States' economic interests could no longer be so subordinated to its political and security alliances. Connally felt the same way.

WHEN CONNALLY ENTERED the Treasury, many who had served David Kennedy, his predecessor, feared for their jobs. But Connally brought with him just two people, one to handle his legal affairs and one to do public relations. In a small staff meeting of top Treasury officials just after he arrived, Connally said, "I'll always back you, but if I don't get your loyalty in return I'll cut your balls off."

Connally proceeded to prepare for his confirmation hearings. "When I took over as Secretary of the Treasury," he wrote years later, "I did so with feelings of trepidation. I was not an economist; I had really never studied monetary affairs . . . I knew I would be supporting policies not because they reflected my philosophy but because they were necessary." Nevertheless, he studied diligently, taking home volumes of briefing books every night and on weekends. He sought out retired Federal Reserve chairman William McChesney Martin for hours of tutoring. He did the same with John Petty, the assistant secretary of the treasury for international affairs, who had had extensive experience in international banking at Chase Manhattan Bank before entering government. Petty knew that the Texan had virtually no knowledge of how the international economy worked. Yet, "[a]fter a tutorial or two," said Petty, "I sometimes felt he knew all I did, but he could explain it better." In his first substantive session with Connally, Treasury Undersecretary Paul Volcker briefed him on the upcoming global monetary turbulence. "I remember my sense of relief," Volcker recalled. "I thought we would have difficulty communicating about the mysteries of the monetary system. We did not."

As Connally was digging in, the top Treasury staff continued to emphasize that a critical priority would be to address an international monetary crisis on the horizon. Sometime in early February, Connally was passing by Volcker's office and casually walked in. Volcker had in his hand a thick policy document, double spaced, with lots of handwriting in the margins. When Connally asked what he was holding, Volcker replied that it was a rough draft of a memo relating to the upcoming potential crisis in global finance but that it wasn't quite ready for him. Connally said he couldn't wait, grabbed it out of Volcker's fist, and left.

CONNALLY WAS CONFIRMED by the Senate on the afternoon of Monday, February 8, 1971, by voice vote. For the next seven months, the new treasury secretary would proceed to shake up the substance and style of America's foreign economic policy. He would be seen as a nationalistic bully on what had formerly been the calm and collegial playing fields of international finance. No one who knew him would have been surprised. John Connally's view of his job was clear from what he said to a group of outside economic experts whom he summoned to Treasury shortly after the Camp David weekend. "It's simple," he told the group of distinguished economists that included Richard Cooper, Henry Wallich, and Fred Bergsten, "I want to screw the foreigners before they screw us."

Paul A. Volcker Jr.

Next to President Nixon, the person most important to John Connally was Paul Volcker, forty-four, his undersecretary for monetary affairs. The two men were the critical negotiators for everything to do with the dollar. That they became such a team was remarkable because they were also a study in contrasts.

Connally had no background in international finance and little interest in its history or the details of how the system worked. He was more interested in trade than in finance. He was an ardent nationalist willing to use protectionism, as necessary, to achieve his broader aims. He was also an extraordinary politician who loved nothing better than bludgeoning his opposition to fall in line with his ideas.

Volcker, for his part, knew the international monetary framework inside and out: he understood how and why it had evolved, its policy framework, its strengths and its shortcomings—even its back-office operations. Economic historian William Greider wrote, "Volcker's training was so extensive that he resembled a civil servant from the British system, a career man who stays in government while the elected officials come and go, who accumulates great influence because he knows every issue more thoroughly than the politicians whom he advises." Additionally, Volcker had formed close associations with leaders on Wall Street and with foreign financial officials in Western Europe and Japan. He believed that increased cooperation across borders was necessary for economic progress and opposed raising trade barriers as a solution to America's difficulties. He was the embodiment of a public servant and apolitical official. He smoked cheap cigars, dressed in rumpled suits that were often shiny as a result of too much wear, and preferred dinner in run-down Chinese restaurants to posh dinner parties. Greider wrote that Connally once threatened to fire him if Volcker didn't get a haircut and buy a new suit.

Despite their differences, Connally and Volcker needed each other, respected each other, and meshed beautifully. One knew how to deal with the president and Congress and how to conduct high-stakes negotiations with major political figures around the world. The other knew the intricacies of what they were negotiating about and the impact that changes in financial arrangements would have on the U.S. and world economies.

Volcker would emerge from the looming crisis as central to the fundamental changes in the global arrangements underpinning the dollar. It was Volcker who headed up the key policy reviews and wrote the pivotal policy options papers. It was he who traveled the world incessantly to orchestrate two major changes in the value of the dollar. Among executive branch officials, it was Volcker on whom congressional experts most relied for their understanding of what the administration wanted to do, and why.

Volcker had one other set of qualities that stood out: unimpeachable integrity. In his dealings within the administration, in his relations with Congress, in his interactions with finance ministries and central bankers, he gave no sense that his goals were other than those he firmly believed were rooted in the national interest—which, in his view, embodied the need for a vibrant global financial system. He reflected no particular ideology. He listened carefully, asked many questions, but at the same time, he was deeply respected as a tough, unflappable negotiator.

A decade after he joined the Nixon administration, Volcker would become chairman of the Federal Reserve and eventually one of the most revered public officials of his generation. In the years 1969–71, however, he was relatively unknown.

BORN IN 1927, Volcker grew up in a family steeped in public service. His father was the town manager of Teaneck, New Jersey, who rescued the town from financial crisis after the 1929 crash and oversaw its administration for twenty years. Volcker tried to enlist in the army in 1945, but he was rejected because of his height—nearly six feet, seven inches. He then went on to Princeton, where he wrote his college thesis on "The Problems of Federal Reserve Policy Since World War II," concluding that the Fed had failed to contain postwar inflation in part because it lacked clear criteria for policy. With a stint at the London School of Economics, he went on to receive a master's degree from the Harvard Graduate School of Public Administration, where he focused on economics but stopped short of a PhD.

His first job in 1952 was at the Federal Reserve Bank of New York, where he worked for vice president and research director Robert Roosa. When Roosa became the JFK administration's undersecretary for monetary affairs in 1961, he asked Volcker to leave Chase and join him, first as an advisor and then as his deputy. It was a formative learning experience, for Roosa was both a great intellect and a formidable financial technician. Roosa was also a passionate supporter of the dollar–gold link and of international financial cooperation. From 1965 to 1969, Volcker ran Chase Manhattan Bank's forward planning, working closely with the chairman, David Rockefeller, among America's most prominent international business leaders.

On Wednesday, December 11, 1968, after Nixon nominated David Kennedy as treasury secretary, Kennedy selected Volcker as his undersecretary for monetary affairs, the same job Roosa had held under JFK. It was one of the government's most powerful positions in the economic arena, for aside from that of secretary of the treasury itself, it was the only job that encompassed both domestic and international affairs insofar as the management of the dollar was concerned. Because the dollar was influenced by interest rates, fiscal policy, trade policy, and taxation, Volcker's remit was exceptionally broad. Given that the dollar played a role in foreign policy, Volcker would also be working with the State Department and the National Security Council. Another advantage for Volcker was that David Kennedy, his mild-mannered boss during Nixon's first two years, felt he had his hands full with other issues and was happy to cede the limelight to Volcker on international monetary policy. In addition, the other agencies of government were limited in their perspective and lacked the credibility that Treasury had in the markets with the business community and with foreign finance ministries. The exception was the independent Federal Reserve, to which virtually everyone listened carefully as well. Volcker was respected in the Fed and understood the importance of the two agencies working as closely together as possible.

Prior to Nixon's being elected, Volcker, a Democrat, had not been a fan of the new president, particularly resenting how he had smeared two-time Democratic presidential candidate Adlai

Stevenson in 1952 and 1956 with a Communist label. But Volcker respected Nixon's extensive foreign policy experience, and he felt that his own passion for public service made the time ripe for trying to fashion a new policy for the dollar. He returned to Washington from Chase with an acute understanding of international finance and the dollar, including all that had transpired in the quarter century since 1944, and with experience in both the public and private sectors.

SHORTLY AFTER HE assumed office on Monday, January 20, 1969, Volcker was sitting in his office at Treasury, from where he could see the inauguration parade moving down Pennsylvania Avenue. As he was watching the procession, an envelope from the White House was delivered, signed by National Security Advisor Henry Kissinger. In it was a memorandum announcing a permanent working group on international monetary policy, to be chaired by Volcker. (In fact, this was a successor to a similar group in the Johnson administration.) The group was directed to conduct a study and recommend policies to the new administration, reporting to the National Security Council. Volcker was pleased to be given the responsibility, but he objected to reporting to Kissinger rather than to his boss, David Kennedy, fearing that the NSC channel would make foreign policy considerations too dominant in policies regarding the dollar. He walked down to Kennedy's office and suggested that the new treasury secretary resist the usurpation of his authority. But judging that his boss was not about to take on Kissinger, Volcker decided to ignore the reporting lines to the NSC and never heard any more about them.

Volcker would drive and control what would be known throughout the administration as the "Volcker Group." It would become the central intergovernmental committee that provided analyses of issues, immediate and long term. In terms of management style, Volcker was an enigma. Although he had a broad and complicated mandate to oversee both international and domestic policies relating to the dollar, he had only a skeleton staff and instead relied on the assistant secretaries overseeing the bureaucracies of both the international and domestic parts of the Treasury. This meant that it was in his own head that everything came together, for it was often just he who had a complete picture of the analysis and the policy implications. This was the way he liked it, given that he was the ultimate close-to-the-chest possessor of information. He would ask lots of questions, propose hypothetical arguments, and hammer out alternative courses of action, but he rarely provided an opinion as to policy prescriptions. One of Volcker's deputies, Bruce MacLaury, described how Volcker worked. "[He] mulled things over in his own mind and even members of his inner group didn't know what his views were. There were books of options, but only Volcker knew what the policy was."

It's not that he lacked strong views. Years later, in an oral history, Volcker talked about his most fundamental belief about the dollar as he was assuming office. He harkened back to the time when he worked for Roosa in the early 1960s and "the $35 gold price was sacred—a matter of international commitment and credibility—and you couldn't talk about changing the exchange rate of the dollar." In the same interview, he recalled that in 1969 a rumor was circulating that Arthur Burns, then a chief advisor to Nixon (and soon to become chairman of the Federal Reserve), favored a devaluation of the dollar. That action, Volcker said, "was anathema to me." Volcker's biographer also describes his views as he joined the Nixon administration and as recorded in Volcker's personal notes at the time. "Price stability belongs to the social contract," Volcker believed. "We give government the right to print money because we trust elected officials not to abuse that right, not to debase that currency by inflating. Foreigners hold our dollars because they trust our pledge that these dollars are equivalent to gold. And trust is everything."

Volcker understood all the rationales for fixed exchange rates and their opposite, floating rates. In favoring the former, Volcker was implicitly saying that the idea of all rates being set against one another with little room for divergence, except as a last resort in dire emergencies, was so important that governments should orchestrate spending, taxing, and interest rate measures in order to keep the supply and demand of their currency stable. Volcker believed that money needed an anchor, some price that was a fixed point, and that an exchange rate tied to gold served that purpose. He was convinced that fixed currencies imparted a necessary discipline on national policy makers, who otherwise would too easily resort to lax fiscal and monetary policies that would lead to inflation. Those who supported floating rates believed that currencies should not be regulated, but they had their eye on something more political in the real world. If your currency could float, you would inevitably prioritize domestic policy matters over the stability of the international currency system.

In some ways, Volcker was in a time warp. It wasn't so much that fixed rates didn't have their attraction, but that they were increasingly a political liability. The critical issue was this: the world was becoming financially interdependent, yet it was organized into sovereign states that wanted the maximum political autonomy. How to reconcile these two forces—globalization and sovereignty—was at the heart of the tension around exchange rates. Governments wanted the maximum discretion, yet the more linked they became by finance, the less control over prices they could have. Fixed rates implied less freedom for governments; floating rates meant more. It wasn't that floating rates had no impact on the U.S. economy, but the effect would be slower to unfold and more diffuse than a major change in interest rates or fiscal policy. Allowing rates to fluctuate would avoid the necessity of making the major political decisions that came with devaluations or revaluations.

For much of the first quarter century after the war, the United States had the capacity to give priority to both international and domestic issues. Volcker wanted a return to those halcyon days. He thought Washington should gear its fiscal and monetary policies to uphold fixed rates, which was what the Western European and Japanese governments wanted the United States to do, too. After all, they were worried about the flood of dollars entering the world economy precisely because U.S. policies were so lax. But most of Washington was moving in the opposite direction, for Nixon felt everything should be done to keep growth going and unemployment low. Over the next few years, Volcker would adjust his understanding as to what was politically possible and necessary. He would also be the person who did the most to negotiate new currency arrangements culminating in floating rates. The rapid evolution of his thinking began in the aftermath of the Camp David weekend.

FOLLOWING FROM KISSINGER'S directive of Monday, January 20, 1969, Volcker brought together high-level representatives of all the key departments, including representatives of State, the Federal Reserve, the CEA, and NSC staff. Their work built on extensive efforts of previous administrations to come to grips with two facts.

The first was that the international monetary system needed fundamental reform. In truth, the Bretton Woods arrangements worked as intended only for a few years. Right after the war, with the exception of the United States, other countries' currencies were too weak for them to be exchanged for their counterparts. No one wanted anything but dollars, and everyone needed dollars for trade. As a result, there was a major dollar shortage. By the end of the 1950s, however, there were too many dollars abroad. In other words, the world went from a dollar shortage to a dollar glut. By the 1970s, excess dollars flooded the world economy. One objective of monetary reform was for the IMF to create a new international asset to substitute for dollars in the future, especially when it came to international reserves. The new asset, as mentioned earlier, would be called "special drawing rights," or SDRs. A huge need arose to lend more money to developing countries, which were becoming bigger players in international finance. The idea was to expand the lending capability of the IMF itself.

The second big problem the Volcker Group addressed was that the United States didn't have enough gold to exchange gold for dollars at the rate of $35 an ounce. During the Kennedy and Johnson administrations, the problem was papered over by a number of ad hoc measures—special tax provisions that discouraged foreigners from taking dollars out of the United States; controls on U.S. bank lending abroad; limits on U.S. multinational companies' investments in Western Europe. Volcker himself knew these controls were symptomatic of the United States' unwillingness to use its budget and interest rates to improve the balance of payments. Also, this web of constraints grew to proportions that made it clear that the U.S. balance-of-payments picture was being artificially manipulated to look better than it would have if capital were allowed to flow freely.

The Volcker Group looked at both problems, but it was the second one, the question of the dollar, that held the potential for a major crisis on the U.S. doorstep. The group spent between February and June 1969 on writing a forty-eight-page, double-spaced document, which was forwarded to President Nixon on Monday,

June 23, 1969. It was discussed in the White House Cabinet Room with President Nixon and with Secretary David Kennedy, Secretary of State William Rogers, Council of Economic Advisers chairman Paul McCracken, counselor to the president Arthur Burns, and NSC advisor Henry Kissinger all attending.

Volcker did the presentation using a series of flip charts. The report itself was the most detailed and comprehensive effort to present policy options for the dollar and the international monetary system in all the Nixon years and would serve as the basis for subsequent reports that were written up through the weekend of August 13–15, 1971. It began with an alarming description of the setting in mid-1969. A brief summary:

- The international monetary system is under great strain. There have been repeated currency crises in the past decade that have undermined confidence in it, and more crises could be under way. (Volcker was referring to problems that had surrounded several countries' currencies, most significantly in Great Britain, France, and West Germany.)

- The United States cannot make good on its commitment to exchange gold for dollars at the fixed price of $35 per ounce if all foreign central banks bring their dollars to the Treasury. U.S. gold reserves are about $11.2 billion, while foreign official holdings are hovering around $40 billion.

- A risk exists that Bretton Woods could break down if everyone wanted gold for their excess dollars and that this could jeopardize American international leadership across all fields.

- Foreign governments are concerned that the United States cannot get its balance of payments under control. They are worried that the inflation in America is leading to a worsening of this balance of payments, which in turn is pushing dollars abroad and then contributing to inflation in their countries.

• The policies that Washington follows now will be central to its domestic and foreign policy, as well as political alignments around the world. Among the challenges is the possible emergence of major trading blocs of countries competing with the United States, most immediately in Western Europe.

• Nothing is more urgent than the United States' getting its inflation under control. "Lack of confidence in relative U.S. price stability (and thus in the future value of the dollar internationally) would for all practical purposes make it impossible to negotiate" on a smooth trajectory. The result would be a series of abrupt emergency measures leading potentially to great instability in global markets.

Volcker then proceeded to outline four options.

First, the United States could "attempt to preserve the present arrangements more or less intact." Volcker quickly rejected this course, implying that, unless significant changes were made, the existing system would self-destruct. Such destabilization could lead to extensive controls on capital and a dangerous level of protectionism, and it would deeply undermine U.S. alliances around the world.

Alternatively, Washington could initiate "a series of multilateral negotiations pointing toward a fundamental but evolutionary change in the existing system." This would entail: encouraging use of the new international asset called SDRs that the IMF had created as international reserves to supplement gold and the dollar; enlarging the IMF's lending programs so that countries could have more time to make fixes in their economy so they wouldn't have to devalue; pressing countries such as West Germany and Japan, both of which had large trade surpluses and were racking up larger and larger financial reserves, to revalue their currencies; finding a way to allow for some additional flexibility (beyond the existing permissible range of 1 percent) for currencies to fluctuate around a fixed parity within defined limits; pushing hard on the trade front for the dismantling of foreign barriers to U.S. exporters; and negotiating better agreements for foreign countries to offset U.S. defense expenditures abroad.

Volcker explained that this "evolutionary" option involved a stepped-up level of responsibility on the part of U.S. allies, but if successful, it could "retain for an indefinite period a major role for the dollar and monetary leadership for the United States." He also admitted that progress could be slow and difficult and it "may not bring change soon enough to forestall the breakdown of the system."

Third, the United States could suspend its commitment of exchanging gold for dollars. This could be a shock treatment for other countries to come to the negotiating table to realign their currencies, open up their economies to more imports from the United States, and better share defense-spending burdens. Or it could be the only response if other countries began a rush to convert their dollars into gold. The risks were unknown reactions and consequences in the international economy and great divisiveness with U.S. allies. Suspending gold conversion could begin a new era of mercantilism, controls over capital and trade, and blocs of countries using exchange rates as economic weapons.

Fourth, the United States could devalue the dollar by raising the price of gold. Thus, instead of pricing gold at $35 an ounce, it would be priced at, say, $38 an ounce, making each dollar worth less in terms of gold. While a cheaper dollar would enhance U.S.

competitiveness, it was not clear how such a devaluation could be made to happen in the absence of a negotiation. The reason was that all currencies were fixed to the dollar at a specified rate, and nothing would stop other countries from devaluing along with the United States and thereby nullifying any advantage the United States hoped to attain. Another factor militating against devaluation was that countries such as West Germany, the United Kingdom, and Japan, which had amassed dollars due to the U.S. commitment to maintain the dollar–gold link, would be left high and dry with dollar reserves that were now valued only by the free market and not gold, as Washington had promised. Just as bad, those countries that mined gold—especially the USSR—would be beneficiaries of U.S. actions. Volcker felt that the "losers" would demand U.S. compensation, making this option even more difficult to contemplate.

Volcker's recommendation was to pursue the second alternative—negotiate a fundamental and comprehensive set of new arrangements in an evolutionary way. If that effort failed to produce results, he raised the possibility that the United States could resort to closing its "gold window" temporarily in order to force other countries to negotiate with it.

To be sure, this wasn't the first time a suspension of the dollar–gold commitment had been raised in internal discussions in the Treasury or in policy papers. But Volcker's June 23 report was the first time the issue was presented to such a high level in the U.S. government. No evidence exists that anyone favored this suspension-of-gold-sales option; on the contrary, it was seen as the ultimate weapon, one not to be used unless there was no other choice.

Volcker also asked for specific presidential guidance on this crucial question: The American stock of gold reserves had been declining. What was the minimal level the United States should preserve? At what point should the administration shut off any conversion from dollars to gold? The current level of U.S. gold reserves cited in Volcker's report was $11.2 billion, of which $1 billion was legally necessary to cover commitments to the IMF. Volcker proposed that the United States not tolerate a level of gold reserves below $8 billion. Beyond that, the administration would have to close the gold window, Volcker suggested.

This much was clear from the presentation: The dollar and the entire Bretton Woods framework were in trouble, and the adoption of a full-fledged U.S. strategy was urgent. The global monetary system was unpredictable and fragile, and no one could rule out another crisis lurking around the corner. All the policy alternatives that Volcker discussed were surrounded by incalculable uncertainties. No one could know how other governments would react, and no one could anticipate what the markets would do.

All that said, throughout the U.S. government a consensus existed that the burden on the United States of managing international monetary arrangements had to be more equitably shared with the Western Europeans and Japanese. The big question was how to push other countries along in directions they had thus far resisted without rupturing political ties and without creating a financial crisis.

At the end of the presentation, everyone awaited Nixon's reaction. After all, it would be hard to envision a bigger or potentially more explosive set of issues than Volcker had just presented. The president's response was short and noncommittal. Nixon said something like, "Good job, and keep me posted on where we stand." Secretary Kennedy approached Volcker and said, "At least he didn't say 'no'" to the minimum floor of $8 billion in gold reserves. (Eventually, going below $10 billion was considered a red line, but that wasn't apparent until the lead-up to the Camp David weekend.) Volcker replied, "I guess we have a policy by default."

In later years, Volcker himself wrote that a change in the gold price—that is, a devaluation of the dollar—"[w]as included among the other options, as far as I was concerned, more for the sake of completeness than any sympathy for the idea." In addition, Volcker himself may have been less than enthusiastic about the viability of his recommendation for the multilateral, evolutionary approach than he let on. In fact, Volcker felt that an exchange rate realignment was necessary and urgent, but he didn't see how it actually could be negotiated. There were just too many con-flicting interests among the United States, Europe, and Japan. In addition, if the United States threatened that it would devalue, everyone else could do the same thing. At the same time, a public give-and-take with other governments would undoubtedly create havoc in the financial markets as the disarray among govern-ments played out in the media. He later acknowledged that he was basically playing for time until the U.S. balance of payments turned around. In my view, he was probably thinking something like this: None of the options is to my liking or really feasible. I am hoping for a near miracle—for the United States to discipline itself by keeping interest rates high enough and budget deficits low enough in order to slay inflation, for Washington to focus on competitiveness, for trade negotiators to genuinely open foreign markets, for defense ministries to put up more money—for *some-thing*, so that we can maintain the only policy that makes sense for us and the world, and that is the status-quo fixed exchange rates, albeit with a devalued dollar in terms of gold, and with much more flexibility for rates to fluctuate around the parities.

Even the views of Nixon's top advisors were not aligned. Trea-sury Secretary Kennedy backed Volcker, but Kennedy was not a strong figure in the administration. George Shultz, who wielded great influence on the economic team, favored doing away with fixed exchange rates altogether. So did CEA chairman Paul Mc-Cracken. Arthur Burns, at the time counselor to Nixon (and not yet chairman of the Fed), was unclear about where he stood, having once voiced a preference for a devaluation but having been vague about the depth of his conviction. Volcker was not sure that his report had changed anyone's mind.

In time, it became clear that Nixon didn't trust Volcker. He felt that the treasury undersecretary cared more about international economic relations and global financial stability than he did about U.S. domestic welfare. The president believed Volcker wasn't dedicated enough to the issues that would most support Nixon's reelection, such as boosting U.S. employment. Remarkably, Volcker didn't suffer from being on the outs with the president. That was because Connally depended on him and respected him for his knowledge and connections in international financial circles, and Connally therefore shielded and supported him in their efforts to reach monetary agreements.

Arthur F. Burns

As chairman of the Federal Reserve Board, Arthur Burns was not formally part of the administration. By law, the Fed enjoyed considerable independence to pursue its policies and reported only to Congress. Its freedom not to kowtow to the administration was assisted by the fact that its key executives—called governors—were confirmed by Congress for fourteen-year terms. By virtue of his powerful and unique position, Wall Street and financial officials around the world listened carefully to every word Burns uttered. After all, he was one of the country's most eminent economists, with experience in academia, independent research institutions, public service, and extensive relationships with business leaders and government officials. Indeed, any public dissension by him regarding the sweeping program Nixon was contemplating at the Camp David meeting would have created a level of confusion and controversy in U.S. and global markets that would have jeopardized public acceptance of any plan put forward.

To most people who have followed the history of the U.S. economy and the Federal Reserve in particular, Burns is remembered as the man who was unable to contain the inflation of the early '70s, an inflation that grew to double digits later in that decade and that ravaged the country. But during the first three years of the Nixon administration, none of this had yet happened.

ARTHUR BURNSEIG WAS born in 1904 in what is today Ukraine and immigrated with his Jewish parents to the United States at the age of ten. The family settled in Bayonne, New Jersey, about twelve miles outside New York City, where Burns's father earned a living as a house painter. In grade school, a teacher shortened Arthur's surname from "Burnseig" to "Burns." He earned a scholarship to Columbia University, working his way through school as a postal clerk, waiter, theater usher, dishwasher, oil tanker mess boy, and salesman, and received his doctorate in 1934. During his studies he became a protégé of Professor Wesley Mitchell, America's foremost expert on the ups and downs of the U.S. economy. In 1946 the two economists coauthored a book on business cycles that in the late '60s was still considered the bible on the subject. Along the way Burns joined the prestigious National Bureau of Economic Research, and in subsequent years became its research director, president, and honorary chairman. His expertise was in subjects such as inflation, unemployment, and methods of smoothing out the ups and downs of the national economy. Until he became Fed chairman, he had much less interest in the international economy.

BURNS MET DWIGHT Eisenhower in the early 1950s when he was a professor at Columbia and the general headed up the university. After taking office in 1953, President Eisenhower brought Burns to the White House as chairman of the White House Council of Economic Advisers, where he remained for three years. It was then that Burns developed a close relationship with Vice President Richard Nixon. In November 1960, when Nixon was running for president against John F. Kennedy, Burns warned Nixon that a national economic slowdown was in store and that it could jeopardize his election prospects. Burns suggested that Nixon press the Eisenhower administration to stimulate the economy immediately. Nixon tried but failed to persuade his colleagues to lean on the Fed to lower interest rates or to accelerate federal spending. Burns's prediction of a recession came true, and Nixon was convinced that he lost to JFK because of a souring economy. He never forgot Burns's advice, nor the lesson that a recession was the political kiss of death for an incumbent.

In Nixon's 1960 campaign against JFK, Burns was a visible member of a support group for the vice president called "Scholars for Nixon and Lodge." Having had a taste of Washington, he gradually lost interest in the academic life and immersed himself in corporate boards, high-level commissions, and other business and policy groups. It was through his National Bureau of Economic Research activities, in particular, that he built up extensive associations with business leaders and former government officials to an unusual extent for an academic economist.

BURNS HAD AN Old World courteousness. His white hair was neatly parted down the middle. He wore rimless glasses and was almost never without a pipe. He spoke slowly and was noted for his patient, even ponderous answers to questions. "He would clean, fill, light, relight, empty and refill his pipe several times during the same conversation," wrote Wyatt Wells, his biographer. According to the *Wall Street Journal*'s John Pierson, "[Burns] takes so long to answer a question between puffs on his ever-present pipe that TV interviewers prefer not to have him on shows." Said *BusinessWeek*, "His brand of humor is described as 'dry' or 'non-existent,' depending on whom is being consulted."

There was a tough, even mean dimension to Burns's personality as well. Comparing him to William McChesney Martin, Burns's predecessor, a New York banker said, "I've never heard anyone mention diplomacy, tact, humility or any of the battery of virtues that served Martin so well, where Arthur is concerned." Burns had little compunction about critiquing someone in public, especially if the subject was economics. For example, he regularly humiliated a particular governor of the Fed who seemed unprepared. He would ask him a question and then bore in. "By the time Burns finished, the staff was choking with laughter and the poor man looked like a fool," wrote biographer Wells. That said, Burns did take an interest in training his staff. He often acted as mentor, pushing them to develop the most advanced statistics and the most sophisticated economic models.

IN HIS 1968 run for president, Nixon selected Burns not only as his chief economic advisor but also as overseer of the staff that wrote position papers on a broad range of public policy issues. Once Nixon was installed in the White House, Burns became counselor to the president and the only White House advisor with cabinet rank, a distinction that afforded him easy access to the president. Of all Nixon's advisors and cabinet members, Burns was the oldest and most experienced.

As counselor to the president, Burns was in charge of more than twenty post-election task forces that recommended new policies to the incoming administration. He acted as talent scout, too. Among his recruits were two key members of Nixon's economic team who would be instrumental at Camp David: George Shultz, who became secretary of labor and subsequently director of the Office of Management and Budget, and Paul McCracken, who was appointed chairman of the CEA. As the administration began its work, Burns was asked to coordinate virtually all domestic affairs, to question new proposals coming from the departments, to prepare new legislation, and to formulate instructions from the White House to the departments about what the White House expected of them. Burns sat in on meetings of the cabinet, on new councils set up for issues such as urban affairs, and on National Security Council meetings dealing with global economic issues.

Soon after he became president, Nixon promised Burns that when William McChesney Martin's tenure as Fed chairman was over in January 1970, Burns could have the job. On Friday, October 17, 1969, Nixon announced his intention to nominate Burns for the Fed chairmanship. A few days later, the president took the occasion to talk to Burns about his upcoming position. Nixon wasted no time demonstrating that the independence of the Fed meant little to him and that he wanted Burns to be a cooperative member of the Nixon team, coordinating Fed policy with Nixon's broader economic goals.

> *Nixon:* My relations with the Fed will be different than they were with Bill Martin there. He was always six months late doing anything. I'm counting on you, Arthur, to keep us out of a recession. (Nixon meant that Martin was much too slow to lower interest rates, which Nixon wanted in order to avoid recession.)

> *Burns, lighting his pipe:* Yes, Mr. President, I don't like to be late.

Nixon: The Fed and the money supply are more important than anything the Bureau of the Budget does.

Burns nods.

Nixon: Arthur, I want you to come over and see me privately anytime . . . I know there's a myth of the autonomous Fed . . . and when you go up for confirmation some senator may ask you about your friendship with the president. Appearances are going to be important, so you can call Ehrlichman and get messages to me, and he'll call you back.

BURNS ENTERED THE Fed at a time of sharp differences of opinion between economists of the Kennedy and Johnson administrations and the emerging group of economists labeled the "Chicago School," led by the high-profile Milton Friedman. The former included such men as former CEA chairman Walter Heller, who believed in managing the economy with fiscal and monetary policies, constantly adjusting them as economic conditions changed. In contrast, Milton Friedman and his followers believed that fine-tuning the economy was impossible and that the only way to curb inflation was for the Fed to expand the money supply in a slow but steady fashion without modifying policy with every up and down of the economy.

Burns also faced an immediate challenge. In early 1970, one year after Nixon became president, both inflation and unemployment were rising, a situation that was eventually called stagflation. Policy makers had not seen this lethal combination before and didn't know how to handle it. In the past, if inflation flared, the economy became overheated and jobs plentiful. Or, if conditions were characterized by high unemployment, economic growth was likely slow and price rises subdued. When both inflation and unemployment existed at the same time, however, policy makers faced a seemingly intractable dilemma. Whichever course they took caused an offsetting problem. Suppose the target was restraining inflation. If Burns tightened interest rates in order to slow the economy and reduce price pressures, he would be risking a recession and higher unemployment. Or suppose the goal was to decrease unemployment. If Burns eased rates in order to improve the job situation, then inflation would flare up. This trade-off between inflation and unemployment was captured in what was called the "Phillips curve," a theory found in most economic textbooks.

Critical international factors had to be considered, too. Here again, solving one problem could create another. If Burns raised interest rates to levels that existed abroad, capital would likely flow into the United States from other countries to take advantage of higher returns. That would help reduce America's overall balance-of-payments deficit and mop up excess dollars from abroad. Yet higher rates could strengthen the dollar, slow exports, and accelerate imports. The converse was true, also. If Burns lowered interest rates relative to those of U.S. trading partners, money would flow out and the balance of payments would deteriorate. The resulting large dollar outflows from the United States would lower foreign confidence in America's ability to manage its balance of payments and put more pressure on holders of dollars abroad to exchange their greenbacks for gold.

Bottom line: from the standpoint of both politics and economics, all policy choices had some negative powerful implications.

BURNS BROUGHT AN eclectic set of beliefs about general economic policy to the Fed. He was not easily identified with the theories of John Maynard Keynes or Keynes's opposite, the free-market advocate Milton Friedman. The new Fed chairman was not terribly interested in theoretical analysis, seeing instead each phase of the business cycle, the booms and contractions of the economy, as embodying a unique set of circumstances. He didn't believe you could fine-tune the economy, as the economists for JFK and LBJ did; nor did he believe it was as simple as maintaining a steady increase in money supply, as the conservative economists were advocating. For Burns, the economy was too complex for any one theory and too dependent on business confidence, the roots of which were not easily defined. "The argument between the Friedmanites and the Keynesians is a false argument," Burns once said. "It's an argument about how well this or that group of economists can forecast the future. They cannot do so, and thank God they can't. If they could, a government could perpetuate itself in power."

Every Fed chairman before him and every conventional economist of the time would have said that the Fed plays a critical role in controlling inflation. But Burns wasn't always convinced. In the late 1960s and early '70s, he saw inflation not as a problem of too rapid an expansion of the money supply but as one of a tug-of-war between big business and big labor unions, both pushing up wages and prices in their bargaining with one another. The phenomenon became known as "cost-push inflation." In Burns's view, the answer to that problem wasn't for the Fed to raise interest rates and create a recession, nor for the administration to rein in spending, which could have the same effect. For him, the answer would not be found in conventional monetary and fiscal policy. Instead, he believed the government would have to exert more pressure over wages and prices. Presidents Kennedy and Johnson had experimented with voluntary wage and price caps, but Burns would eventually move even further toward actual controls. This notion was anathema to Richard Nixon and virtually all conservative Republicans, and so Burns didn't reveal his thinking at first, but by late 1970, his convictions caused a major rift between him and the president.

WHEN IT CAME to international monetary matters, a subject Burns had not studied intensively in the way he had researched the nature of business cycles, his views were erratic.

In October 1968, when he was coordinating the development of policy perspectives for Nixon's campaign, Burns made a trip through Western Europe and concluded that the United States ought to devalue the dollar by raising the price of gold, and do this in a multilateral agreement. (Raising the price of gold meant that it would take more dollars to equal an ounce of the precious metal, rendering every dollar worth less in terms of gold—hence, a devaluation.) It is not clear how strongly Burns held this view, but in any event, it did not appear in the post-election task force reports. When he entered the Federal Reserve, Burns became enmeshed in the web of international fora for central bankers and de facto leaders representing the world's central currency. In that position, he had to play the role of stabilizer of the global monetary regime, which would have meant discouraging talk of dollar devaluation.

In the spring of 1971, however, he met with the French minister of economy and finance, Valéry Giscard d'Estaing, and was taken by the latter's proposal for a series of bilateral meetings, followed by a bigger multilateral one, to devalue the dollar and revalue other currencies. It was a proposal similar to one he had embraced before the election. He mentioned Giscard d'Estaing's idea to President Nixon that same day and recorded in his diary that Nixon asked him to pursue it. There is no evidence that Burns followed up.

When the August 1971 Camp David meeting arrived three months later, Burns first opposed closing the gold window, the only person there to do so. But before the weekend was over, he gave up on that position, too.

REGARDING THE INDEPENDENCE of the Fed, the old adage that what you believe depends on where you sit seemed apt. Less than one month into the Nixon administration, while Burns was acting as counselor, for example, Fed chairman William Mc-Chesney Martin raised interest rates in a bid to slow the red-hot economy and tamp down inflationary pressures. Nixon was upset. At a cabinet meeting, Burns said, "Eisenhower liked to talk about the independence of the Federal Reserve. They [meaning the Fed] begin to believe it. Let's not make the mistake and talk about the independence of the Fed." But when Burns assumed the Fed chairman's job, his views changed. At his confirmation hearings on December 18, 1969, he had this exchange with Sen. William Proxmire:

Proxmire: I am particularly concerned with the tendency of the Federal Reserve Board—if it does not have strong leadership and independent leadership—to become a kind of handmaiden to the Treasury.

Burns: You have expressed my views perfectly . . . I assure you that I will always do what I think is best and what is sound for the Nation, without regard to any considerations of partisan or political or extraneous advantage.

Of course, Burns was not about to have a fight with Proxmire at his confirmation hearings, but in fact, for much of 1970 and

1971, Burns did act independently—certainly in the eyes of the much-annoyed Nixon White House. However, by the time of the Camp David meeting in August 1971 and throughout 1972, an election year, it is hard to conclude anything other than that the Fed chair caved to Nixon's pressure.

BURNS'S DIARIES REVEAL the agony of his relationship with the president. He clearly wanted to please Nixon and demonstrate his loyalty, even as he pursued policies that Nixon and his team resented. On Monday, March 8, 1971, for example, he described a White House meeting he attended in which the president berated his economic team for not putting a more positive shine on the administration's economic policies. "I am convinced the President would do anything to be reelected," Burns wrote. "The harassing of the Fed by the President and his pusillanimous staff will continue and even intensify. Fortunately, I am no longer sure whether the President fully knows this, I am still his best friend. By standing firm, I will serve the economy—and thereby the President—best."

Later that month, on Sunday, March 21, Burns reflected on the negative stories he believed others in the administration were planting in the media that exaggerated policy differences between him and the president. In response, he requested a private meeting with Nixon at which, Burns later wrote, he told the president "that his friendship is one of the three that has counted most in my life and that I wanted to keep it if I possibly could . . . that there was never the slightest conflict between my doing what was right for the economy and my doing what served the political interests of Richard Nixon . . . that if a conflict ever arose between these objectives, I would not lose a minute in informing [him] and seeking a solution together."

Four months later, on Thursday, July 8, 1971, Nixon demanded that Burns adhere more closely to administration policy. Burns confided to his diary of "the brutality of Nixon's language . . . I watched his face, as he spoke, with a feeling of dismay; for his features became twisted, and what I saw was uncontrolled cruelty." Burns then replied to the president that in the event of a rift between him and the administration's policies, "It would be most unfortunate. I just want you to know that I have always been your true friend, and I am now and expect to remain so." Then a realization came over him, and he recorded this in his diary: "Now I knew that I would be accepted in the future only if I suppressed my will and yielded completely—even though if I was wrong [under] law and morality—to his will."

Burns's desire to be Nixon's friend may have been heightened by his disdain for those around the president who he felt didn't understand the nature of inflation or how to deal with it. He described William McChesney Martin, Nixon's first Fed chairman, as a "pathetic slob." He wrote that Paul McCracken, the CEA chairman whom Burns had recruited, was "obsessed with his own status." Haldeman and Ehrlichman were described as having "amoral leanings." George Shultz was both a "pernicious and stultifying force," but later he is a "loyal and devoted servant." John Connally lacked capacity for leadership. Kissinger had an "egomaniacal approach." Paul Volcker was "an indecisive man full of flaws and anxieties."

In sending Burns to the Fed, Nixon undoubtedly wanted someone he thought he could control. It was therefore no surprise that the president was furious at Burns for pursuing policies he thought were undercutting his efforts and made the administration seem to be in disarray. The tension between Nixon and Burns was further exacerbated by the fact that several senior officials in the administration, most notably George Shultz, held strong views contrary to Burns's policies and often berated him behind his back, sometimes to the press on a not-for-attribution basis.

THE PRESIDENT'S PRESSURE to influence Burns was on display from the outset of Burns's tenure at the Fed. On Saturday, January 31, 1970, in a morning ceremony in the East Room of the White House to commemorate his move to the Fed, Burns received a standing ovation when he was introduced. "You see, Dr. Burns," Nixon said to the assembled crowd, "that is a standing vote of appreciation for lower interest rates and more money." Nixon then made some additional remarks, a bit tongue-in-cheek, but not entirely. "As all of you know, the Federal Reserve is independent . . . I respect that independence. On the other hand, I do have the opportunity as President to convey my views to the Chairman . . . I have some very strong views on some of these economic matters and I can assure you that I will convey them privately and strongly to Dr. Burns . . . I respect his independence. However, I hope that independently he will conclude that my views are the ones that should be followed." At the same ceremony, Burns said, "My duties at the Federal Reserve Board, I think, can be described in one sentence: to do what I can to help protect the integrity of the dollar and to help foster a stable prosperity for the nation." The tension was evident. Nixon wanted easy money, which would lead to increased employment but could lead to inflation and the erosion of the dollar's value. Burns was advocating the opposite. Anyone parsing both men's remarks would have seen the seeds of a blossoming conflict.

The pressure campaign on Burns began hardly two months into his tenure as Fed chairman. On Friday, March 20, 1970, Nixon used John Ehrlichman, his chief domestic policy advisor, as an intermediary. In a phone call to Burns, Ehrlichman relayed Nixon's thoughts that "Responsibility for recession is directly on the Fed. It's a very tight situation . . . They [the Fed] must free up construction money [for housing] now or it's too late . . . The president will take on the Fed publicly." Ehrlichman was intimating that Nixon would be pressing Burns through a number of powerful intermediaries ranging from senators and congressmen to Wall Street executives. Nixon was even prepared to publicly criticize Burns's policies. By the standards of the early 1970s, such a breakdown in relations between the much-admired Arthur Burns on the part of an experienced president would have seriously rattled markets.

George Shultz egged on Nixon's growing hostility toward Burns, telling the president, "Arthur has a way of holding the money supply hostage to [fiscal policy]." He accused the Fed chairman of saying that unless Nixon reduced the budget deficit, he, Burns, wouldn't lower interest rates. Burns was in fact saying that when it came to fiscal policy, the onus for managing the economy should not be on the Federal Reserve only, but also on the administration and Congress. If monetary and fiscal policy were both loose, then higher inflation would be inevitable. If Nixon wanted Burns to stimulate the economy, the federal deficit would have to be restrained. Shultz's characterization of Burns helped further inflame Nixon. At a cabinet meeting on Monday, April 3, 1970, the president slammed his hand on the table. "When we get through," the president said, "this Fed won't be independent if it's the only thing I do."

After the disappointing midterm elections in November 1970, Nixon didn't want to take any chances with the presidential election two years hence. At all costs, he would prevent a slowdown in growth and an increase in unemployment. Nixon saw Burns as the key actor in averting this situation and wanted him to expand the money supply by lowering interest rates, but by then Burns was on a different track.

BURNS'S BELIEF THAT government controls on prices and wages—and not on monetary or fiscal policy—was the only solution to combatting inflation became the theme of a series of speeches he made about the U.S. economy in 1970. He started with suggestions for voluntary measures that business and labor could take to restrain wage and price increases, but he gradually concluded that only mandatory regulations would suffice. While Nixon loathed such controls, the list of government officials who supported some kind of direct controls on wages and prices was nevertheless expanding. Several members of Nixon's cabinet, within the confines of internal deliberations, favored them, including commerce secretary Maurice Stans, housing and urban development secretary George Romney, and transportation secretary John Volpe. In his private moments, Paul McCracken, who had publicly opposed controls, had growing doubts about his own position, too. The business community sympathized with Burns as well. At a convention on Monday, October 19, the Business Council, the most prestigious group of CEOs, pushed voluntary controls, not out of national altruism but with the hope that the administration would lean on the unions to stop pursuing such large settlements. The Democratic-led Congress was also squarely on Burns's side. On Thursday, August 13, 1970, it passed the Economic Stabilization Act, giving the president the authority to freeze wages, salaries, and rents. The one-page bill came in the form of a rider to a law relating to national security, the Defense Production Act, that made it impossible for Nixon to veto. It was designed to embarrass the president by making it clear that he wasn't doing all he could to kill inflation and that now he had the legal tools to do more. The act was renewed several months later. On both occasions, Nixon declared that he had no intention of using the authority.

In contrast to Burns, Nixon abhorred any kind of across-the-board intervention in the private sector, and he grew incensed that Burns was publicly beating the drum for wage and price controls. Nixon believed that Burns had been eliciting support from different constituencies and boxing him in. Nixon told Ehrlichman, "Burns will get it in the chops." Ehrlichman recalls Nixon yelling, "Is it time to take on the Fed in public? We won't take this. Should we give the Fed a good kick now?"

On Friday, November 20, 1970, Burns had a one-and-three-quarter-hour meeting with the president. "I emphasized time was short, that he would eventually have to adapt [wage and price controls]. That events will force such a move, that chances of success in [the presidential election in 1972] would be best if he moved promptly." Burns recalled that Nixon listened sympathetically. But because he famously avoided confrontation, no one can be sure of what the president actually felt.

Nevertheless, toward the end of 1970, with no progress being made on inflation, Nixon reversed course, or so it seemed, and suggested that some limited wage and price restraint might make sense under certain circumstances. On Friday, December 4, in a speech at the Waldorf Astoria, Nixon made a link between his new flexibility on wage and price controls and a commitment he said Burns had made to lower interest rates and increase the money supply. If there was a link, Burns refused to acknowledge this publicly. To Nixon's great annoyance, Burns gave another speech a few days later, taking an even tougher stand on controls. Still, the media sensed that some kind of agreement was emerging between the two men—namely, that if Nixon took measures on wages and prices, Burns would open the monetary taps. Leonard

Silk of the *New York Times* labeled the public interplay between Nixon and Burns as the "Accord of 1970." *BusinessWeek* agreed, saying, "In effect, Burns was offering the White House a concordant that could develop into a new framework for economic policy: monetary and fiscal policy would both be turned toward rapid expansion, and continued inflation would lead to intensified government intervention in private economic decisions," meaning that Washington would forcefully oppose business and labor efforts to increase prices and wages by imposing controls.

If there had been an implicit agreement between the two men, it did not last. Nixon reverted to his strong opposition against controls, even though rising wages and prices from collective bargaining continued to play havoc with the economy for the first half of 1971. In a message to Congress on Monday, February 1, 1971, the president wrote, "I do not intend to impose wage and price controls which would substitute new, growing and more vexatious problems for the problem of inflation." On the same day, CEA chairman Paul McCracken wrote in his annual report, "Mandatory price and wage controls are undesirable, unnecessary, and probably unworkable." Years later, Herbert Stein, who would succeed McCracken as head of the CEA, reflected on what was happening. "The administration and the economy were engaged in a race," he wrote. "The question was whether the administration's disinflation program would be seen to be succeeding before disappointment with its failures made the demands for [wage and price controls] irresistible."

IN THE LEAD-UP to the Camp David agreement, the tensions between Nixon and Burns over the right approach to inflation would become worse, raising the major question of how they could reconcile each other's positions during the fateful weekend.

George P. Shultz

George Shultz would become one of America's great statesmen of the last third of the twentieth century. By the time he left government in 1989, he had served with distinction as secretary of labor, director of the Office of Management and Budget, secretary of the treasury, and secretary of state. After his government service he wrote prolifically and thoughtfully on all manner of public policy issues.

In December 1968, Shultz was a relatively unknown dean of the University of Chicago Graduate School of Business when Nixon tapped him to become secretary of labor. Shultz quickly rose within the administration as an influential advisor. He was noted for the clarity with which he advanced an argument, for his ability to listen carefully to whomever he was talking to, and for control of his emotions in tense situations. Nixon knew that even if Shultz's views were overruled, as they often were, Shultz, the ultimate team player, would faithfully execute Nixon's policy. He had another side to him as well—that of a fierce conservative partisan who was known to undermine the policies of others behind their backs.

If Connally represented the absence of ideological conviction and an obsession with short-term politics; if Volcker was a superb financial technician with political skills who was trying his best to defend the status quo; and if Burns was a conservative whose policies were, in fact, hard to categorize, Shultz was something altogether different: a true and consistent conservative. Among Nixon's men at Camp David, Shultz foreshadowed more than anyone else the Thatcher-Reagan revolution of extensive deregulation that was less than a decade away.

SHULTZ WAS RAISED in New York City. His father held a PhD in history from Columbia University and worked on the New York Stock Exchange as a resident professor. Shultz graduated from Princeton in 1942, where he played varsity baseball and football, then fought the Japanese as a marine officer from 1942 to 1945. In 1949, he earned a PhD in industrial economics from the Massachusetts Institute of Technology, and he stayed on at MIT to teach from 1949 to 1955. For the next two years, he served as a senior economist on Eisenhower's Council of Economic Advisers, then chaired by Arthur Burns. He left the administration to teach industrial relations at the University of Chicago Graduate School of Business, becoming dean of the school a few years later. Throughout his academic career, he was deeply involved in public policy and was well respected in both Republican and Democratic circles. He served as an arbitrator of labor-management disputes across a broad range of industries—electrical equipment, farm machinery, textiles, chemicals, and food products, for example—as well as on many corporate boards and federal government task forces established in the JFK-LBJ years.

SHULTZ CHAIRED A task force during Nixon's 1968 presidential campaign that concluded that price controls were ineffective and counterproductive. In December 1968, at Arthur Burns's suggestion, Nixon interviewed Shultz for the position of secretary of labor. The president-elect wanted to make sure that the Chicago dean agreed with the philosophy of keeping the federal government out of labor-management collective bargaining, in contrast to the heavier involvement by the two preceding Democratic administrations. With that affirmed, the deal was sealed.

Nixon valued the Chicago professor's versatility. Shultz quickly became involved in a broad range of issues critical to the administration, many going beyond the traditional remit of the Department of Labor. He oversaw the settlement of the strike of the International Longshoremen's Association that had closed down ports on the East Coast and the Gulf of Mexico without federal intervention. He orchestrated the settlement of a major postal strike. He devised measures to deal with discrimination in the workplace. He spearheaded efforts to deal with school desegregation in the South. He evaluated options to replace quotas on oil imports, and he assessed the economic consequences of the United States' turning over the burden of fighting in Vietnam to the government in Saigon. Shultz was at heart a rational analyst and someone who could bridge differences among contending parties. At no time did he see himself as a politician. "I don't think that the President looks at me as a great font of wisdom about how to get elected," he once said. "If he did, I'd be alarmed."

On Wednesday, July 1, 1970, Nixon expanded the role of the Bureau of the Budget (BOB) to become the Office of Management and Budget (OMB), an enlarged organization with powers that went well beyond budgets and would extend to every corner of the executive branch. The president appointed Shultz to the job and, signaling its importance, gave him an office in the White House—something former budget directors never had. From that perch, Shultz would meet with Nixon once or twice each day and assumed operational and planning responsibilities for virtually every major domestic issue. He took over the 7:30 a.m. White House meeting Ehrlichman had been running. Of Shultz, *BusinessWeek* wrote, "He is one of the President's three or four top confidants. He is a pivotal man in the drive to restructure the executive branch . . . He has now emerged as the chief architect of the administration's program for economic recovery." Said James Naughton in the *New York Times*, "Every White House eventually produces an individual who is relied upon so heavily that he becomes, in effect, Assistant President. The current example is Mr. Shultz."

SHULTZ WAS VALUABLE to Nixon for another reason: he had good relations with George Meany, president of the powerful AFL-CIO union. Big Labor was suspicious of any Republican administration, but Nixon was facing a particularly restive union movement intent on raising wages by tough collective bargaining and, if necessary, long and expensive strikes. Labor had also become a protectionist force, in contrast to its free-trade position throughout the 1950s and most of the '60s, and threatened to undermine Nixon's desire to expand trade. Shultz's relationship with Meany was thus crucial; indeed, when Shultz was appointed director of OMB, Meany, not known for his compliments of administration officials, issued this statement: "George Shultz has served with distinction as chief of the Department of Labor. He has fully deserved the confidence that American workers, their unions and the AFL-CIO have placed in him."

SHULTZ WAS DESCRIBED by *BusinessWeek* as "a shirt-sleeve type, pipe-smoking, and somewhat rumpled." His temperament was one of a negotiator and mediator. He spoke softly, rarely raising his voice, and at cabinet meetings others often had to lean forward to hear him. He believed it was worth taking the time to get everyone's views before making a decision. Paul Volcker recalled of Shultz, "Time and again he would work with almost inhuman patience to bring a group into agreement upon a decision all could support, at times submerging his own preferences." Shultz once described his way of operating: "With the changing political winds, one who sets sail directly toward his goal would never get there. The skill lies in the tacking." Still, Shultz found no virtue in routinely splitting the difference to win agreement. He was fond of saying, "He who walks in the middle gets hit from both sides."

Shultz impressed Nixon with his ability to translate ideas into action, so much so that Nixon considered him "the only real knowledgeable economist" in the administration. Speechwriter William Safire judged Shultz the fiercest advocate of the free-market economy in the administration. The key to his ideological beliefs was his strong association with the "Chicago School" of economists, the major tenets of which include: the free market is the most efficient way to allocate resources; we should be deeply skeptical about government intervention in the economy; market-based solutions are the best way to approach not just economic but also social problems; and freedom is more important than equality. Shultz was a student, admirer, and close friend of the school's most well-known leader, Milton Friedman, with whom he stayed in close touch while in the administration. Despite the close association, Friedman placed less emphasis on Shultz's ideas than on his character and effectiveness in managing people.

"George is a man of principle, but he is not an ideologue—like I am," Friedman said. "His forte is not in academic analysis but in problem solving."

Shultz was a monetarist, meaning that he believed that the Federal Reserve's ability to control the money supply was the key governing tool for the economy, far more than fiscal policy (which the two previous Democratic administrations had emphasized). He thought the Fed should expand the money supply according to a set formula, and in 1969–71, he wanted that formula to be geared to faster growth. As we have seen, Shultz was especially critical of Arthur Burns during that period for what Shultz considered Burns's failure to expand the money supply at a faster rate.

Shultz disagreed with Burns on another matter: wage and price controls. He objected to what he considered heavy-handed government intervention in the private sector. His opposition to such controls mirrored the president's, but like Nixon, he ultimately tolerated them. In fact, once the decision to institute controls was made, he and his staff became the principal architects of the new system, designing how it would work and be administered.

SHULTZ WAS ALSO an ardent free-trader, believing that all barriers to international commerce should be dropped as low as possible, and he quietly opposed Nixon's textile quotas and any tariffs to slow imports. When it came to the dollar, Shultz wanted to abolish almost all features of Bretton Woods. He would have liked to end the role of gold in backing currencies, abolish fixed rates for currency relationships, and let currencies float against one another. In his view, the market for currencies should be like the market for any commodity; values should be determined in an open market governed by supply and demand. He strongly disagreed with Paul Volcker, who, he thought, correctly, was determined to maintain the essence of the Bretton Woods arrangements. As he did with Burns, Shultz on occasion criticized Volcker to Nixon in the heated lead-up to the Camp David weekend and in the follow-up. (Later, when Shultz succeeded Connally as secretary of treasury in 1972, he became an admirer of Volcker, on whom he came to rely.)

As for Connally, Shultz recognized him as a master politician, incredibly smart and quick, but not a strategist. And he was a great admirer of Henry Kissinger, whose foreign policy remit didn't cross over into Shultz's bailiwick.

NO ONE WOULD have called Shultz a compelling orator. Historian Allen Matusow described him as "stiff as a board." *BusinessWeek* called him "the grayest man in a gray cabinet." Nevertheless, Shultz gave one speech that he himself would cite time and again as capturing the essence of his beliefs. The speech was called "Prescription for Economic Policy—'Steady as You Go,'" delivered on Thursday, April 22, 1971, and it was a plea for a strategy that relied on consistent monetary growth and conservative budgeting. He warned against succumbing to political pressures for excessive stimulation of the economy that existed on both the budgetary and monetary fronts. He was reemphasizing the importance of the policy of gradualism and opposing the policies of Burns, who had abandoned gradualism in favor of wage and price controls. For Shultz, the basics of monetary and fiscal policy still counted, and they should be given a chance to work. Policy makers should be patient. They should not jump to fashionable theories that the structural nature of the economy had changed, because it had not changed all that much.

FOR ALL HIS quiet competence and for all his achievements small and big, Shultz fought many losing battles in the years to come. But Nixon trusted him deeply and wanted him in the room when big decisions were made—and he most always was.

9

Peter G. Peterson

Peter Peterson would become one of the country's most renowned businessmen, statesmen, and financial mandarins, but at the beginning of 1971, he was little known outside corporate America. Nevertheless, he embodied a set of assets many of the others on the Nixon team lacked, especially deep credibility with business leaders. His focus was on the real economy, such as industry and technology, as opposed to just finance and economic policy. More than anyone else on the Nixon team, he grasped the enthusiasm and confidence of the United States' high-technology business sector, including a sense of unlimited possibilities for the country if only it established the right policies. Before and after Camp David, he became a prime link between the corporate world and the upper echelons of policy making.

THE ORIGINS OF Peterson's appointment could be found in late 1970 in the recommendations of the Ash Commission, the same group from which Connally had attracted Nixon's attention. One of the commission's recommendations was to significantly elevate attention to international economic policy, given its increasing importance to the nation's domestic and foreign interests. From a U.S. vantage point, foreign relations with, say, Western Europe, were traditionally more dominated by national security than by trade. Fiscal and monetary policy at home failed to take enough account of international capital flows. This was no surprise, because U.S. reliance on international economic transactions was relatively small; the country was, to a great extent, a major continental nation bordered by two oceans, and was more or less self-sufficient. But the Ash group saw that the situation was changing fast. For Western Europe and Japan, economic relations with the United States were of prime importance in their overall relationships to America. It was a mistake to think that they subordinated these issues to traditional foreign policy challenges relating to, say, military affairs. In addition, the international economy was bearing down on the United States. For example, trade and the dollar were beginning to affect U.S. domestic employment at a time when the latter became a red-hot domestic political problem. For reasons like these, the Ash Commission recommended the establishment of a new group in the White House akin to the National Security Council. It would be called the Council on International Economic Policy (CIEP), chaired by the president, just as the NSC was, and supported by an executive director and staff who could consider all aspects of the U.S. role in the world economy. Like the NSC, the CIEP would coordinate the many agencies (State, Treasury, Commerce, Agriculture, etc.) that, in this case, had a stake in the development of the United States' international economic strategy.

No follow-up to the Ash Commission recommendation for the establishment of the CIEP or its executive director was evident for several months, but in late December 1970, the administration was ready to move ahead. George Shultz, who knew Peter Peterson from the University of Chicago Graduate School of Business, where Peterson had once taught in the evenings, called to ask if he would see the president about leading the new council. Peterson, forty-four at the time, was chairman and CEO of Bell and Howell, then widely known and respected for manufacturing cameras and other audiovisual equipment. One of five children from a Greek immigrant father with a third-grade education, Peterson had attended Kearney State Teachers College for one year, then MIT for another, and then Northwestern, where he graduated in 1947. He started work at a research company while studying marketing in the evenings at the University of Chicago Graduate School of Business, where he earned an MBA. He moved to McCann-Erickson, a large advertising firm, where he pitched and lured clients such as Peter Pan (peanut butter), Rival Packing Co. (dog food), and Swift & Co. (meat packing). Eventually, Peterson joined Bell and Howell as executive vice president for marketing, then rose up the ladder to run the firm.

His experience at Bell and Howell from 1959 to 1971 was particularly formative. During this time, Peterson was immersed in the opportunities and the travails of an American technology company in an increasingly competitive global market. He came to understand the extraordinary possibilities that existed in the camera and video sectors, but he also witnessed firsthand the fierce competition from Japan in consumer electronics, a trend that would soon lead to a more defensive U.S. trade policy. Peterson's response to the challenge was to partner with a Japanese company, Canon Inc., an innovative transaction at a time when few U.S. companies were doing such joint ventures in Asia. His insights from these experiences imbued him with a sense that rapid and continual changes in technology would characterize the evolution of the world economy and that, without a major fight with their up-and-coming rivals, U.S. firms would not enjoy the unrivaled lead they were used to.

NIXON INTERVIEWED PETERSON in early January 1971, laying out his views about the growing importance of global economic trends, including the rise of West Germany and Japan. "Nixon was envisioning the interconnected geo-economic world that globalists talk about today," Peterson recalled. He accepted the job, which encompassed two titles that conferred White House status: executive director of the Council on International Economic Policy and assistant to the president for international economic affairs.

In a White House ceremony appointing Peterson held on Tuesday, January 19, 1971, Nixon said, "He is a man that has been described by his colleagues in the business community as one of the ablest—and some even used the term brilliant—chief executives of his generation." Peterson, who had a dry sense of humor, later recalled his reaction to Nixon's over-the-top compliment. "It was the surest sign so far that I had left the area of quantifiable data and entered the Washington realm of hyperbole and spin."

In his announcement, Nixon explained why Peterson's job was so needed, referring to issues such as trade and energy that were overlapping with foreign and domestic policy. Peterson's mandate would inevitably bring him in direct competition with many cabinet heads, including John Connally and Secretary of State William Rogers, as well as NSC advisor Henry Kissinger. "Pete Peterson is stepping into one of the toughest jobs in Washington," said *Fortune* in an editorial. Indeed, coordinating policy among agencies with different perspectives and different agendas would be an exceptionally challenging mandate—and eventually part of Peterson's undoing.

At the beginning of his tenure, Peterson sent memos to Nixon that reflected key issues he wanted to delve into concerning America's position in the international economy. He indicated that he had been consulting widely both inside the administration and out to put together an agenda. Nixon initially showed warmth and respect for Peterson, who, he at first felt, "had a mind full of penetrating questions about international economic policy." Peterson had arrived as a knee-jerk free-trader, with a strong inclination to allow market forces to determine outcomes. But when he saw how economically powerful West Germany and Japan had become, how much more open U.S. markets were to imports than were its major trading partners, how overvalued the dollar appeared to be, and how unprepared the United States was to deal with the imbalanced state of international trade and finance, he quickly became a proponent of an aggressive policy toward the allies. To him, that meant confronting their trade barriers with U.S. imposition of tariffs and quotas and ramping up Washington's investments in American technology so the United States could maintain its competitive edge. He also supported reforms of the dollar-based international monetary system, although he was careful not to encroach on Connally's turf. Most of all, he captured Nixon's attention by focusing on the decline of U.S. competitiveness and the measures necessary to reverse the nation's deteriorating position. These concepts appealed to Nixon's penchant for big ideas and bold responses.

PETERSON WAS TASKED by the Council on International Economic Policy with preparing a major report on "The United

States in the Changing World Economy." Within a few months, he produced a draft that became an influential document in Washington. His accomplishment was less in producing original content than in assembling massive amounts of information and organizing it into a powerful, clear narrative with easy-to-grasp color graphics—unusual for that time for a government report. Like Connally, Peterson had a talent for digesting reams of figures and extracting patterns. He could assemble advisors who were far more knowledgeable about their respective areas of expertise than he and wring out of them more than they thought they knew. Then he would stitch their contributions together into an intriguing story. He proved to be an exceptional marketer of ideas, adjusting and refining a presentation after each session to make it sharper, more compelling. He was the rare official who habitually took a long view, explaining why it was urgent to act now before the government's policy options were foreclosed.

The "Peterson Report" was a 133-page document packed with statistics and charts. It was billed as a personal document of Peter Peterson, backed up by the staff work of the CIEP, but was never approved by any cabinet-level body. (John Petty, one of Connally's key assistant secretaries, told me in an interview that Treasury was reluctant to clear such a wide-ranging document for fear they would be ceding authority to Peterson.) The report made several key points. Among them:

- The world economy had changed dramatically since the 1950s. The mind-set of Americans and the country's policies had failed to adjust. "How we meet the economic challenge in the 1970s may have more to do with what kind of country we become and indeed what kind of world it is than any other thing we do," the report said.

- The United States had been losing ground to Western Europe and particularly Japan, and unless policies of the United States and its allies changed, U.S. competitiveness would continue to erode. Beyond that was a looming threat from up-and-coming developing countries, too.

- The United States must lead the international effort toward trade and monetary policies in which the global burdens of managing the world economy were better shared with other nations. The United States will have to negotiate with "a clearer, more assertive vision of the national interest . . . We must dispel any 'Marshall Plan psychology' or relatively unconstrained generosity that may remain . . . This is not just a matter of choice but necessity."

- Nevertheless, many of the challenges the United States faced were right at home. The nation must give new urgency to investment in its own competitiveness and in policies that helped workers and communities adjust to changing technologies, import penetration, and the outsourcing of production by multinational companies.

In early April, the report went to Nixon, who was so taken by passages that quantified the positive contribution that exports could make to U.S. job creation that he underlined that part of the report and showed it to visitors for the next several weeks. "Once he grasped the implications of economic nationalism for his domestic purposes," wrote historian Allen Matusow, "Nixon committed wholeheartedly" to the report. He asked that the document be circulated within the administration and that Peterson brief members of Congress and business and labor groups.

On Tuesday, June 29, in extemporaneous remarks before the National Commission on Productivity, Nixon seemed to be reflecting on the report when he said, "It's terribly important that we be #1 economically [in the world], because otherwise we can't be #1 diplomatically or militarily." In late June, Nixon and Connally discussed the importance of a newly aggressive trade policy for winning the support of the labor movement. Connally brought up the possibility of establishing restrictions on imports to gain labor support.

On July 6, the *Wall Street Journal* emphasized Peterson's work on its front page. "Mr. Peterson's graphs and charts, depicting trade trends, are starting to awaken some powerful people here to a growing challenge the U.S. faces in international trade." The article highlighted what it called some "out-of-the-box" ideas that Peterson advanced. These included government subsidies for research and development in selected technologies, a softer approach to antitrust policies that would help combinations of U.S. companies compete internationally, tax breaks for exporters, and a realignment of exchange rates. Peterson also urged consideration of an industrial policy akin to what had emerged in Japan and Western Europe in which government and industry planned together for exploiting markets of the future.

John Connally was a particular booster of Peterson's report, as it unequivocally supported his own aggressive, nationalist leanings. And John Ehrlichman would tell *Fortune* magazine that Peterson's work was "the starting point for dramatic changes in economic policy that Nixon announced on August 15, 1971."

ON MONDAY, JULY 12, 1971, Peterson sent a confidential memo to Nixon revealing his concerns about America's future, going beyond what was in his report. In the memo, Peterson referred to his extensive contacts in the business world and their concern that Washington needed a better handle on where its future competitive advantages lay. Peterson talked about meetings he had had with business leaders, including CEOs from prominent firms such as GE and ITT (International Telephone and Telegraph, then a gigantic international conglomerate), who favored extensive planning for the future. He summarized their views as showing corporate America's lack of confidence in the country's future and a recognition of a need to take their own fate in their hands more than usual. "We have experienced our first smell of defeat (i.e., in Vietnam and international competition)," they said, according to Peterson. He paraphrased other sentiments: "We need to sense and shape a new and exciting future—a new sense of purpose."

Understanding that Nixon prided himself on being a hands-off-business Republican, Peterson, ever the pragmatist, gingerly proposed that the government play a bigger role in selectively investing in the industries of the future, as Japan was doing and as, he warned, the European Community was on the verge of doing. He said no place in government was asking the question "This is where we are . . . this is where we want to be and this is how we get from here to there." Peterson was cautiously suggesting that America couldn't compete effectively unless it engaged in the sort of industrial planning in ordinary industries that occurred in the national defense sector. This line of thinking was badly received by most Republicans around Nixon, as they saw it as government intervention in the economy akin to socialist or Communist states. And while a conservative Democrat such as John Connally, who as a pragmatic governor of Texas had dedicated himself to investment in technology, might have welcomed the idea, his hands were full with other matters at the time. I could find no record of Nixon's response to Peterson's far-reaching planning proposals.

PETERSON WAS MOTIVATED in large part by big ideas, but he also knew that his bureaucratic position was fragile. After all, his mandate was to bridge the jurisdictions of many departments with long-standing missions and bureaucrats who jealously guarded their traditional activities. John Connally, despite his enthusiasm for Peterson's ideas, saw himself and Treasury as having total charge of international economic policy and would prove to be Peterson's biggest nemesis. He simply did not want Peterson sharing the limelight with him. Peterson recalled, "I naively assumed that Connally would have been supportive of my work . . . because he had been a member of the Ash Council [*sic*] that had recommended the creation of the Council on International Economic Policy. It took me a while—longer than it should have—to realize that the position I held posed a threat to Connally's ego and ambitions." Henry Kissinger, who had forged a highly cooperative relationship with Peterson, wrote that "Connally had reduced Peterson to the role of spectator."

Connally wasn't Peterson's only problem. The former Bell and Howell CEO had become a member of the so-called Georgetown set, a group of illustrious former officials in the administrations of JFK and LBJ, plus influential columnists, reporters, and assorted intellectuals from think tanks. He dined with people like Katharine Graham, owner of the *Washington Post*, played tennis with former JFK defense secretary Robert McNamara, and hobnobbed with columnists such as James Reston of the *New York Times*. Cavorting with this crowd earned Peterson a black mark in the administration. Nixon himself resented the Georgetown elite, feeling that they had rejected him ever since he was vice president and, what's more, were out to undermine him. The result was that Nixon and his close aides grew to feel that Peterson was an untrustworthy member of the team.

The Nixon entourage also saw Peterson as an egregious self-promoter who was not above leaking selected information to make himself look more powerful than he may have been. In the same way he could market ideas, Peterson was adept at marketing himself by capitalizing on his relationships with the media, often using his charts and statistics to educate reporters and, not incidentally, implying how important he was to many key decisions. "Nixon's White House entourage saw him as over prone to reach for power, and overzealous in seeking public credit for his accomplishments," wrote Juan Cameron of *Fortune*. Cameron depicted Peterson as "an ambitious loner." He said Peterson violated his own maxim, "Never get yourself stereotyped. If you do, you lose your credibility."

Despite the controversy around Peterson, Nixon would invite him to the August 1971 Camp David meeting and allowed him to be part of the follow-up. While Connally pushed Peterson away, when it came to anything to do with monetary affairs, Nixon asked the former CEO to take charge of textile negotiations with Japan after everyone else had failed. These negotiations were critical to Nixon himself, given his campaign pledges to southern lawmakers that he would rein in Japanese exports. "I don't give a damn what it takes," Nixon told Peterson. "You do whatever is required to [bring it to a close]. Do you understand?"

In the end, Peterson's firsthand experience in the business world, his extensive relationships with CEOs, his marketing savvy, and the breadth of his thinking allowed him to hang on through the end of the critical year of 1971. On Tuesday, February 29, 1972, Nixon made him secretary of commerce. It was a cabinet position, to be sure, but everyone, including Peterson, knew he was being expelled from the inner circle.

Other Players—Paul W. McCracken
and Henry A. Kissinger

While Nixon was the key decision maker, five other men—
Connally, Volcker, Burns, Shultz, and Peterson—would all prove
critical voices when the crisis all had been fearing came to a head.
They were not the only ones who would play significant roles in
shaping the events of the weekend, however.

PAUL MCCRACKEN, THE chairman of Nixon's Council of
Economic Advisers, was an active member of the Nixon brain trust
and influential in setting the stage for the Camp David weekend,
in which he participated. A highly respected economist, he came
from a farming family in the Midwest, graduated with a PhD
from Harvard, and served on the staff of the CEA under President
Eisenhower (just after Burns had departed). McCracken was not
a hard-line conservative. He leaned toward the free-market ideas
of Milton Friedman, but not entirely, and preferred to describe

himself as "Friedmanesque." And like Connally, Volcker, Shultz, and Peterson, he had been involved in projects that spanned both Democratic and Republican administrations. Ideologically, he was something of a transition figure between the Keynesians of the 1960s (economists who advocated highly active fiscal policies) and the more conservative economic ideas of the late 1970s and early '80s that ultimately produced Ronald Reagan. He was a stalwart defender of the administration's policy of gradualism, of which he was a key architect. Always a team player, McCracken was nevertheless capable of changing his mind in fundamental ways. Just before the Camp David weekend, for example, he reluctantly concluded that the wage and price controls he'd once opposed would be necessary. Nixon listened to him, consulted with him, and included him in most of the critical deliberations on domestic and international policies.

ANOTHER INFLUENTIAL FIGURE was National Security Advisor Henry Kissinger. Kissinger, of course, became one of the most influential American diplomats of the twentieth century, but in his first three years in Nixon's service, his reputation, although growing rapidly, was not yet what it would eventually become. He was not directly influential in the actual formulation of international monetary policy. Nor was he present at the Camp David meeting. Nevertheless, he and his staff played three roles that influenced the big decisions made in August 13–15, 1971, and carried out in the last four months of that year.

First and most important, Kissinger, along with Nixon, was the chief intellectual architect for the overall shift in American foreign policy from a position of single-handed dominance over the free world to one in which political and economic power and responsibility would have to be shared. Before coming to

Washington, the NSC advisor was a prominent professor, writer, and consultant who espoused such views, and within the administration, his influence on the president in foreign affairs was unequaled. "The age of the superpowers is now drawing to an end," he wrote shortly before he was appointed by Nixon. "And there must be a conviction that the United States cannot or will not carry all the burdens alone." It fell to Kissinger to manage the implementation of Nixon's foreign policy—including the strategy that served as an umbrella for future negotiations on international economic and financial policies in the aftermath of the Camp David weekend. Wrote David Rothkopf, a historian of U.S. national security policy, "There was no doubt about how foreign policy was being made in the Nixon administration. Nixon drove, Kissinger navigated, and others had the choice to make comments from the back seat or get out of the car."

Kissinger's second role was to recruit and oversee a highly skilled economic staff for the National Security Council, which became a consistent voice within the administration for pointing out the foreign policy implications of U.S. positions on financial matters before and after Camp David. Indeed, Kissinger's staff was involved in many of the Treasury-led discussions in the run-up to the Camp David weekend. Although loaded with talent, it was a very small group. At first it included C. Fred Bergsten, a young, highly skillful economist with extensive knowledge of foreign policy. Bergsten was joined by Robert Hormats, a freshly minted PhD from Tufts. By the summer of 1971, Bergsten was gone, but Hormats attended selected sessions of the Volcker Group, and he maintained strong personal relations with Volcker himself, as well as with Arthur Burns. Hormats also mobilized expert advice for Kissinger on the part of three savvy economists who had studied the relationship between foreign policy and international economic affairs and who had significant government experience in the intersection of both. These included Fred Bergsten, who had moved to the Brookings Institution; Yale University's Richard Cooper, who had been at the State Department in the Johnson administration; and Harvard's Francis Bator, who was on LBJ's NSC staff overseeing transatlantic relations. Altogether, considering that most economists did not have experience with foreign policy, it was an impressive brain trust in and out of government.

Although Bergsten and Hormats bombarded Kissinger with memos keeping him informed of what was happening on the trade and global finance front, and although Kissinger signed many memos to the president on international economic policy that were drafted by his staff, the NSC advisor himself didn't want to play any significant role in international economic policy. His hands were full with other very high-profile international issues. Besides, between his staff and his few key relationships in the administration, he felt he would be alerted to any foreign policy issues deriving from international economic relations that he needed to know.

And third, Kissinger's most important contribution was to bring the negotiations to a harmonious end and to avert a permanent rift among the allies. This was no small feat because in the aftermath of Nixon's abrupt, unilateral decisions over the Camp David weekend, America's allies would emerge shocked and angered.

NEITHER NIXON NOR Kissinger was comfortable with international economic policy. They both saw trade liberalization as a way to strengthen alliances, although Nixon the politician was more sensitive to the politics of import penetration and the need to make exceptions to free-trade policies—or, at least, his career as a politician had caused him to be exposed to this issue more than Kissinger. They both had almost no understanding of international monetary arrangements, other than that in the

1960s, currency crises were a source of instability in the international political landscape. Nixon had at least an intuitive sense that the system was outdated, while Kissinger's interest focused entirely on whether this or any other set of arrangements would strengthen or weaken U.S. alliances.

Kissinger had confidence in Peter Peterson, who kept him in the loop on many urgent matters. He also had a warm mutual relationship with Arthur Burns, perhaps not surprising for two German-Jewish émigrés who had reached the pinnacle of power. And Kissinger held John Connally in great respect. These men were the two giants in Nixon's cabinet. Kissinger was well aware of Nixon's admiration for Connally and the exalted status the president bestowed on him. Accordingly, the NSC advisor was not about to impose himself in Connally's policy bailiwick unless the president specifically asked him to do it—signaling to Kissinger that the issue would be one of indisputable high foreign policy importance. In the event, this is exactly what happened after the Camp David weekend.

Kissinger wasn't at Camp David because he was on his way to Paris for secret negotiations with the North Vietnamese to end the Vietnam War. Would he have otherwise been invited? It is hard to say. Although the decisions at Camp David had extensive foreign policy implications, Nixon may have considered himself the ultimate foreign policy strategist and would have seen his NSC advisor as superfluous. The most likely explanation for why Nixon might not have wanted Kissinger at Camp David, even if he had been available, is that the president knew many knotty foreign policy issues would be involved and that the mere raising of them at this late date would cause delays in decisions Nixon wanted to postpone no longer.

Had Kissinger been there, he likely would have argued for more advanced consultations with the allies in order to contain their surprise and possible bitterness, especially on the eve of planned presidential summits with China and the Soviet Union, for which allied support would be essential. He might have pushed for the establishment of a small group of U.S. officials to define negotiating objectives more clearly and to identify potential collateral damage to alliances. He might have raised a number of delicate foreign policy balancing acts that would be necessary to perform. For example, the challenge of dealing with Western European allies who were in the process of forming a tighter economic community created genuine complications. Kissinger might have asked how Washington should approach the EC members so as not to anger them so much that they would be pushed into forming an economic bloc that would be even more competitive with the United States than the individual allies already were, not to mention more protectionist, too. When it came to Japan, Kissinger might have voiced concerns about not alienating Tokyo, which was already deeply upset with Nixon's announcement made just a month before that he would be going to China—a Cold War enemy on Japan's doorstep and one with deep historical animosity against Tokyo.

Whether Kissinger would have even tried to challenge Nixon and Connally on their international economic strategy at Camp David was highly unlikely. In the first instance, he was agnostic on the subject, except that he did not want to see the allies in disarray. Also, Kissinger was deeply uncomfortable talking in a group on subjects that he could not dominate by virtue of his own knowledge.

IN ANY EVENT, right after the Camp David weekend Kissinger was left with an explosive international political mess to clean up, and his interventions in Washington and abroad would be critical to the eventual outcomes of the weekend's decisions.

III.

THE WEEKEND

The Wolf at the Door

Shortly after the monetary crisis in early May 1971, John Connally attended a major international financial conference in Munich. It was Friday, May 28, and the first time that foreign central bankers had seen the Texan in action. Connally's strident nationalist views had been on full display in Washington, but this summit would allow foreigners to determine how much his positions were designed for domestic political consumption and how much they represented his deeply held beliefs.

For example, about six weeks before the Munich event, overseas officials heard that Connally had talked to the *Washington Post* editorial board. "We can't continue to hold a military, economic and political umbrella over the free world ourselves as we have been doing," he told them. "We need a radical change in our trade position." Referring to Western European trade preferences for their former colonies in the Mediterranean—preferences that put U.S. exporters at a disadvantage because they allowed European products to be imported more cheaply than those from the

United States—Connally told the *Post*, "If that's the way they feel, the United States should pull its Sixth Fleet out of the Mediterranean and let the Europeans arrange for their own defense." In the middle of the Cold War, this threat was incendiary and represented a total about-face from previous U.S. policy, which would not have connected trade and defense policy so directly and crassly. Officials overseas wondered: was Connally bolstering his political credentials at home or was he expressing official policy?

At Munich, they quickly discovered that the treasury secretary pulled no punches. In his speech to the group, Connally identified what he felt were egregious imbalances in burden-sharing among the allies when it came to trade, currencies, and defense costs. The postwar world was over, he strongly implied. "The comfortable assumption that the United States should—in the broader political interests of the free world—be willing to bear disproportionate economic costs does not fit the facts of today . . . No longer can considerations of friendship or need or capacity justify the United States carrying so heavy a share of the common burdens."

Despite his diatribe, however, at the end of his presentation he indicated that the United States had no intention of changing its policy. "The Nixon administration is dedicated to assuring the integrity and maintaining the strength of the dollar. We are not going to devalue," he said. "We are not going to change the price of gold. We are controlling our inflation. We are also stimulating economic growth at a pace which will not begin new inflation." It appeared that Connally was ruling out closing the gold window and that the Bretton Woods monetary system would be maintained as long as the Western Europeans and Japanese did their part on trade liberalization and took on more expenses for the common defense.

The media provided a range of interpretations of what Connally meant. To some, Connally proclaimed the Nixon Doctrine in tough, blunt terms. To others, his remarks portended an inward, protectionist path for the United States. Despite Connally's commitment not to upend the monetary system, a West German official nevertheless told Clyde Farnsworth of the *New York Times* that the speech signaled "the old monetary system was dead, and now the world had to wait for a new one to emerge."

Paul Volcker, who had accompanied his boss to Munich and drafted his speech, was in a state of angst. Before he addressed the forum, Connally had gone over Volcker's text and toughened the language. But after seeing Connally's changes, even the more traditional Volcker thought it was dangerous to make an unequivocal pledge to maintain the greenback's value, because his studies had shown that devaluation was a real possibility. Just before Connally went onstage, Volcker questioned him about the advisability of ruling out that eventuality. "That's my unalterable position today," Connally replied. "I don't know what it will be this summer."

VOLCKER'S RESERVATIONS ABOUT Connally's commitment not to devalue the dollar were based on all the analytical work he had been doing in the past several months, including his updated contingency plans for a U.S. response to a major global monetary crisis, an effort, as we have seen, begun in the first half of 1969. Now Volcker had been more secretive than ever, confiding only in a few trusted Treasury aides for fear that leaks of confidential U.S. plans could lead to panic in the markets. Volcker's secrecy was also a reflection of deep-seated rivalries between Treasury and State, a contest heightened by

Connally's desire to maintain total control over all aspects of economic policy, domestic and international.

Aside from its tone of imminent danger, Volcker's contingency plan differed from what he had presented to Nixon in June 1969, when he offered broad options. Now he was zeroing in on what must be done immediately. In fact, Volcker's position had taken a dramatic turn as he had come to conclude that the dollar could no longer be saved with only incremental steps. He jettisoned the recommendation he put forth to Nixon in 1969 for a multilateral negotiation to settle all issues. Now he believed that only by closing the gold window and stopping all exchanges of gold for dollars could the United States make clear to its allies that negotiations were urgent to shore up the framework for currencies and modernize international monetary arrangements. "I came to believe that sooner or later we would have to suspend our promise to convert dollars into gold as a means to an end: the only way to force exchange rate realignment and serious [long-term reform]," Volcker later wrote. "We needed to find the right time to take the initiative."

The treasury undersecretary was pessimistic for several reasons. For one thing, he felt that America's competitive challenges, as reflected in the deteriorating balance of merchandise trade, were just too great to be handled by more domestic investment, enhanced R&D, and similar conventional measures that would have effects only over several years. A devaluation would have to be part of the package of initiatives. Volcker also would have preferred higher U.S. interest rates in order to keep capital from leaving the United States and attracting even more of it from abroad. That would strengthen the dollar and demonstrate to other countries that Washington was serious about reducing its balance-of-payments deficits. Nevertheless, he understood the political reality that

Nixon would push for just the opposite—that is, lower rates—in order to keep the economy growing and reduce unemployment.

Volcker's aggressive revised plan was called "Contingency Planning: Options for the International Monetary Problem," a sixty-three-page document that went through many iterations up until mid-August. It was one of these drafts that Connally pulled out of Volcker's hands when he came through his office sometime in early 1971. Several versions of the plan are in Volcker's files, but together they paint a dire picture of a looming international financial crisis that could erupt as soon as later that year. "Pressures on the international monetary system are rapidly building once again," Volcker wrote. "Reasonably foreseeable events—possibly in a matter of weeks—could set off strong speculation and strain one or more of the basic elements of the present fixed exchange rate system, [pushing] convertible currency arrangements to the breaking point." As part of his warning, Volcker enumerated all the risks for the United States and the world in not acting to deal with an imminent crisis, including the chances that a monetary crisis would lead to increasingly high trade barriers at home and abroad and to a 1930s-type protectionist global response.

The constantly evolving contingency plan talked about both the monetary and nonmonetary objectives that the United States should pursue. In the case of the dollar, Volcker identified several goals. The first was for West Germany and Japan to revalue their currencies in order to make the U.S. dollar cheaper in world markets. The second was to allow for more flexibility for exchange rates to go up and down than was permitted under the Bretton Woods arrangements. That would avoid the need for so many formal currency devaluations and revaluations, all of which would be politically traumatic for the governments involved and would cause major disturbances in markets. Next, Volcker was proposing a phase-out of gold as the major source of reserves, to be replaced by special drawing rights (SDRs), the IMF-created currency that governments could use to settle accounts with one another (but which could not be used commercially). The United States also wanted to see Western Europe and Japan open up their markets for more imports and share the costs of the common defense more equitably.

By May 8, about three weeks before Connally's Munich speech, Volcker's plan was substantially fleshed out. It included the following specific provisions: (1) The United States should suspend gold sales to get the Europeans and Japanese to understand that Washington was not posturing; (2) Washington should provide investment incentives for U.S. business and for the auto industry in particular, to get the economy growing; (3) the U.S. government should implement budget cutbacks to demonstrate it was getting its own economic house in order; and (4) Washington should impose wage and price guidelines or an outright freeze on wages or prices to keep prices down.

The wage and price restraints would become the part of the package that captivated most public attention in the country, as it amounted to a far-reaching intervention in the U.S. economy that affected virtually all businesses and workers. The rationale for controlling wages and prices was clear: nothing else seemed to be working to hold down inflation. At this stage, however, this radical plan was backed only by Connally and Burns. It had yet to be sold to George Shultz and Paul McCracken, let alone Nixon.

Volcker reluctantly felt action on wages and prices was essential for international reasons, too. If the dollar became cheaper relative to other currencies, as Volcker hoped, then imports would become more expensive, putting upward pressure on prices throughout the U.S. economy. Also, somewhere along the line, a temporary 10 percent across-the-board tariff was added to the Volcker plan as a device to force others to negotiate with the United States. It appears Connally asked Volcker to insert this provision even though Volcker said it felt too protectionist for him. Because such a tariff, like a devaluation, could also be inflationary, further increasing the price of imports, it was even more important to levy controls on wages and prices.

Volcker was all too conscious of the risks of whatever course the United States decided on. "Embarking on this course," he wrote in his plan, "we should do so only with our eyes wide open as to the risk of monetary disturbance both to the prestige of the United States and to the effective operation of the international monetary system."

WHILE VOLCKER WAS preoccupied with all aspects of contingency planning, CEA chair Paul McCracken was weighing in with the president, too. On Wednesday, June 2, 1971, for example, he reflected on the May monetary crisis. In a memo to the president, he expressed support for much more flexibility in exchange rates, implying that he favored floating rates. He also suggested that the president take a major initiative to urge the IMF to revamp existing monetary arrangements. This implied that Nixon should not move unilaterally but instead work jointly with other countries, a sentiment at odds with the thinking of Connally and Volcker.

On Monday, June 14, in preparation for a meeting with the president, Connally, Burns, and Shultz, McCracken wrote Nixon another memo expressing his concern about the administration's inability to control inflation, an arena in which his advice, as chairman of the CEA, was supposed to be front and center. To date, McCracken and Shultz had adamantly opposed across-the-board measures to restrain wages and prices. Among their reasons was a practical one: they believed that such controls had never worked in peacetime and that they would end up doing even more damage to the economy by distorting all workings of the market. Nixon shared these views. Still, as a measure of his apprehension about the current dynamic of high prices, expansive wage settlements, plus a moribund economy, McCracken ended his memo: "With no joy, I have concluded that we must even be prepared at a suitable time to invoke wage and price controls."

DURING THE WEEKEND of Saturday, June 26, 1971, the president and his economic advisors met at Camp David to discuss the deteriorating economy. It was a fractious meeting, with a variety of views being presented and no clear path forward offered. Immediately following the meeting, Nixon called his cabinet together again in the White House to castigate them for not all being on the same policy wavelength. In his diary, Haldeman paraphrased the president: "We have a plan, we will follow it, we have confidence in it . . . If you can't follow the rule, or if you can't get along with the Administration's decisions, then get out."

Nixon then said he wanted a single economic policy message from his administration. From then on, John Connally would thus be the only economic spokesman, he announced. Shultz was not happy. Burns, head of an independent Fed that was not legally part of the administration, would surely object to not having his own direct channel to Nixon. Nixon told Ehrlichman, "Just tell Arthur to report to Connally. [Tell him] the President won't see him."

Connally was then told to brief the press, which he did in characteristically clear and blunt terms. There would be no price controls, he said. There would be no tax cut. There would be no increase in federal spending. The bottom line: it would be, as Shultz wanted, steady as she goes. Connally implied that the policies were working; they would just take time. He was, of course, being disingenuous, for he knew that the administration would soon announce dramatic policy departures that would be directly counter to what he had said at Munich and what he was saying now in the aftermath of the just-concluded Camp David meeting.

Over the next twenty-eight days, with the global economic situation deteriorating and the administration still in a muddle about what to do, the necessity for a weekend summit to hammer out a dramatic and decisive plan increased. Historian Allen Matusow called the period the "death watch for Bretton Woods."

ON THURSDAY, JULY 15, Nixon stunned the nation with an evening television address announcing that Henry Kissinger had returned from a secret trip to China and that he, Nixon, would be going to Beijing to seek to normalize relations between the two countries. The next day, the president briefed congressional leaders on the spectacular diplomatic breakthrough. "I found that for every one who expressed support for that foreign policy initiative, at least twice as many used the opportunity to express concern about our domestic policies and to urge new actions to deal with the problems of unemployment and inflation," recalled the president in his memoirs. After the meeting, Connally told Nixon that they had better act fast. "If we don't propose a responsible new [economic] program," the treasury secretary warned, "Congress will have an irresponsible one on your desk within a month." The president's and his advisors' anxieties were raised when a Harris Poll conducted in July showed that 73 percent of respondents had an unfavorable view of the economic performance of the administration, and a Gallup Poll indicated that half of all Americans backed a wage and price freeze. Nixon authorized Connally to consult privately with Shultz, McCracken, Burns, and Peterson and put together a comprehensive plan.

Connally already had the elements of one; it was Paul Volcker's continually updated contingency options document. But the ideas had to be sold to the president, and assuming he bought it, the follow-up would require that all members of the administration act as a unified team. Nixon, Connally, and Shultz thought that if an outlier in the group existed, it would be Arthur Burns.

THE PRESIDENT HAD long been concerned with Arthur Burns's going rogue. Since assuming the Fed chairmanship, Burns had failed to hold interest rates low enough to make the economy grow faster, antagonizing Nixon. The president was also worried about Burns's public advocacy of a wage- and price-control policy that had more teeth than anything Nixon felt he could endorse.

On Friday, July 23, the rift between the White House and Burns opened wide when Burns testified before Congress's Joint Economic Committee. The Fed chairman expressed skepticism about the progress that the administration had made in combatting inflation and unemployment, and voiced concern about America's deteriorating balance-of-payments position. He underlined his view that the combination of rising inflation and rising unemployment at the same time meant "the rules of economics are not working quite the way they used to," and he doubled down on a recommendation he had made in previous months for a wage and price board. In saying that the answers couldn't be found in fiscal and monetary policy, but only in wage and price controls, Burns was espousing a radical theory for most of the economics profession. "It was like an explorer discovering that in the vicinity of the North Pole, his compass spun around and pointed south-southwest instead of north," wrote the *New York Times*'s Leonard Silk.

Nixon was furious with Burns and determined to teach him a lesson. Suddenly, press articles began to appear undermining the Fed chairman. First, it was rumored in the media that the president was considering enlarging the board of the Fed—in a sense, packing it with Nixon supporters and thereby undercutting Burns's capacity to get his way on policy. Newspapers and broadcasters also reported that the administration was considering eliminating the Fed's independence and making it report to the Treasury. And finally, and more personally, "unidentified sources" said that Burns had asked for a pay raise, which was seen as a hypocritical move given that he had been demanding wage and price restraints for the entire economy. "This week the deep hostility between the Administration and the chairman of the Fed broke out in the open," wrote *BusinessWeek*. "Clearly, officials are trying to force Burns back in line."

Richard Janssen and Albert Hunt wrote in the *Wall Street Journal*, "[It] could become a confrontation of historic proportions." Janssen and Hunt speculated that the administration's goals could be to stop Burns's crusade for wage and price controls, to ease him out of his position for someone more compliant, or to make the Fed a scapegoat in the event the economy was failing while the 1972 presidential election campaign was under way. Their article indicated that both Connally and McCracken were deeply opposed to treating Burns this way, but not necessarily Shultz or others on the White House staff who were known to be at odds with the Fed chairman.

A few days later, Nixon believed he had done enough to put Burns on notice and recognized the danger of open warfare with him. In a wide-ranging press conference, he pretended to be above the fray and said that Burns had taken an "unfair shot." He clarified that Burns hadn't asked for a pay raise for himself, only for his successors. He also praised Burns's performance and said that the only difference between the two was whether to impose mandatory wage and price controls. Nixon himself called Burns to patch things up and may have backed down a bit on wage and price controls. Burns wrote in his diary, "On August 4, the President opened the door to a new wage and price policy—not widely to be sure, but he at least indicated that he may take another look. On August 5, when we talked on the phone, he stated—on his own initiative . . . that he might surprise me one of these days."

THE PRESIDENT HAD become increasingly worried over the summer that the United States was losing its competitiveness. Some of his anxiety could be seen on Tuesday, July 6, 1971, when he spoke in Kansas City, Missouri, about the major changes in the world and America's need to adjust to them. He then reiterated the Nixon Doctrine not only as it related to the geopolitical situation, but also as it described how Washington needed to change the U.S. approach to the global economy. "The United States, as compared with that position we found ourselves in immediately after World War II, has a challenge such as we did not even dream of," he said. Using a poker analogy, he went on to explain that in that earlier period, America had had all the chips and was happy to pass some around. "But now," he explained, "we face a situation where . . . other potential economic powers have the capacity . . . to challenge us on every front." Nixon went on to talk about other civilizations that did not adjust their policies to new realities. "I think of what happened to Greece and Rome," he said, "and, as you see, what is left—only the pillars."

About two weeks later, Nixon met with Paul McCracken to discuss the competitiveness of the American economy and the magnitude of what he faced with inflation, unemployment, balance-of-payments deficits, and gyrations in the currency markets, all of which were related. Following that meeting, Nixon called Connally to ask him to gather his thoughts on the currency situation and to approach it "with an open mind." Presidents don't always have to give detailed orders; sometimes an understated suggestion is all that is needed.

On Monday, July 26, Nixon spoke to Peter Peterson in the Oval Office about how to deal with the dollar. Nixon also wanted to discuss how tightly to restrict the group of people considering the issues, in the same confidential way that the China initiative had been handled. Peterson asked for permission to pull together a few top advisors and develop a bold economic plan. Although Peterson knew of the work that Volcker had been doing, he felt it was too narrow in scope, not taking into account such issues as the foreign policy consequences of economic decisions or the precise determination of where they wanted to come out in negotiations. Peterson then saw the limitations of his ambition when Nixon told him to go ahead, but that the key person had to be Connally, because the treasury secretary understood politics.

The next day, Nixon met with Peterson and Connally together. Connally told Nixon that a financial crisis was likely and that with the United States' gold reserves at dangerously low levels, a response could not wait until the 1972 election. "It's going to take some drastic action on your part regardless," Connally said. He promised he would soon present Nixon with a plan, which would be extremely bold. It would be Connally's plan, not Peterson's, and not an interagency plan. It would also essentially be a slightly polished version of what Volcker had painstakingly been working on for over two years.

On July 28, 1971, the *Wall Street Journal* reported of rumors in Paris that the United States might abandon its commitment to convert dollars into gold at the $35 price. On the same day, the *New York Times*'s Clyde Farnsworth reported, "There are growing expectations in Europe that Washington may declare a gold embargo." The rumors were linked to the continuous weakness of the dollar in private markets and the drain on gold. The Treasury referred everyone to Connally's statement at Munich that the United States would not devalue and that it would not change the price of gold (which was another way of saying the same thing). Over the next few days, the media kept accentuating negative developments in the United States, such as widening budget and trade deficits. Of particular concern everywhere was that the merchandise trade surplus that was $2.7 billion in 1970 had become a deficit by the summer of '71 of $600 million. It was the first merchandise deficit since 1893 and a disturbing indication that the country's basic competitiveness in goods was in trouble.

ON MONDAY, AUGUST 2, a few days before Nixon's conciliatory call to Burns, a four-hour meeting between Nixon and Connally, Shultz, and Haldeman in the Oval Office changed everything. Connally arrived with a thick briefing book prepared by Volcker and his team. This was essentially Volcker's latest contingency plan, updated as of July 27. Section "A" contained a bogus plan, in the event that the report leaked. Section "B" was blank, in order to confuse anyone not authorized to have it from trying to make sense of the book. The real plan was in section "C."

In terms of domestic policy, the plan proposed:

- A resumption of the investment tax credit of between 5 and 7 percent to stimulate investment and growth;

- Removal of the excise tax on autos, to stimulate car sales;

- Budget cuts to show a modicum of fiscal responsibility and please die-hard conservatives; and

- A freeze on wages and prices for ninety days (with no indication of whether that would be the end of controls).

The international dimensions would include:

- Embargoing all sales of gold—meaning closing the gold window—which would free up the dollar from its anchor, not to mention shock other countries into earnest negotiations;

- Floating the dollar against other currencies with the near certainty that the dollar would float downward compared to the West German mark and the Japanese yen;

- Eventually renegotiating the fixed parities such that the West German and Japanese currencies would be revalued and the dollar would be permanently devalued compared to the stronger currencies;

- Returning to fixed rates with the newly realigned values;

- Allowing some more flexibility for currencies to fluctuate around the newly fixed rates; and

- Imposing a 10–15 percent across-the-board surtax on all imports until the currency negotiations were concluded to Washington's satisfaction; in other words, imposing temporary import barriers as negotiating leverage.

Connally explained that the domestic and international parts of the package had to be announced together, as they reinforced one another. In short, some of the package would be inflationary, such as a cheaper dollar (which makes imports more expensive) and the import tax (which also raises prices of imports), but constructed the right way, a wage and price freeze would, by definition, put a ceiling on rising prices. The economy would get moving again, the global monetary system would be stabilized, and the United States would become more competitive because of a depreciating currency. Most important—and this was key to Connally's mind-set—the whole was greater than the sum of the parts in political terms, because the package would be seen as big and dramatic, and it would show that Nixon had thought of everything. And the total package contained something for every constituency—business, labor, exporters, consumers, investors, liberals, conservatives.

Nixon was amazed. He recalled years later, "I knew that by nature, he [Connally] always favored the 'big play,' so I would have expected he would recommend something bold. But even I was not prepared for the actions he proposed. He urged, in effect, total war on all economic fronts." After his initial presentation, Connally told Nixon, "I am not sure this program will work. But I am sure that anything else would *not* work."

Connally pitched the plan as a boon for the 1972 presidential election. "This ought to show people you are aware of the problems in the foreign and domestic fields and that you have the courage to face up to it," he said. "[You] take a position before you are forced to." Connally boasted, "It'll be as big a coup as your China thing, no question."

Shultz left the meeting after about an hour. On his way out, he warned the group to proceed carefully when it came to wage and price controls, which he strongly opposed. And he was worried about the gold and currency dimensions of the plan, because he didn't want the president's reelection prospects undercut by the charge that Nixon had devalued the dollar, a move that could be interpreted as a political retreat. Nixon raised few questions about the specific details of the plan. At one point, he asked whether the fate of the dollar could be negotiated, as opposed to the United States acting unilaterally. Connally responded that the publicity surrounding any multilateral discussion about the value of currencies would result in everyone's dumping the dollar, as it would be widely assumed that Washington was engineering a cheaper currency to rectify the chronic merchandise trade deficits (as a cheaper dollar would make U.S. exports cheaper in world markets and imports into the United States more expensive).

Nixon became preoccupied with how the plan would be presented to the public, whether it was necessary to announce the domestic and international parts together, and whether everything should be postponed until Congress reconvened on September 7. At one point, his anxiety became palpable, as he wondered aloud whether they should wait until after the 1972 election to do the whole thing.

Connally saw the president's wavering and pressed back diplomatically but forcefully, explaining that an early date was necessary to prevent being preempted by out-of-control market events or by Congress, or both. He highlighted the precarious gold position the United States was in, saying that the $10 billion reserve level was about to be breached and predicting that it was possible the United States would lose another $3.6 billion by year's end. (Recall that Volcker had previously recommended that the rock-bottom-acceptable level of gold reserves was $8 billion. At some point over the past several months, the acceptable-minimum figure was raised by $2 billion.)

After the meeting, Haldeman wrote in his diary, "This becomes a rather momentous decision, and it would be interesting to see what develops . . . Connally is pushing hard for it. Shultz will put some brakes on, but I'm not sure he'll be able to be effective."

BETWEEN AUGUST 2 and August 12, several additional meetings would take place, but now the elements of the proposal were on the table and remained essentially constant. Other than the president, Connally and Shultz were the only two true insiders in the process, although Burns was let in a few days later. In the background was Paul Volcker, whose contingency plans were the only comprehensive documentation on which Connally's integrated package of proposals rested. McCracken and Peterson were on the outside looking in. Notably absent altogether from the insider group was anyone from the foreign policy community.

Time and again, Connally's role was to persuade the president to move fast and always to keep the entire package in focus. Shultz was a patient and persistent provocateur, asking questions and raising strategic issues, such as what should happen after the freeze and what should be the ultimate goal with regard to the kind of monetary rules and regulations the United States wanted once the value of the dollar was reset. Neither Connally nor Shultz hesitated to flatter the president, underlining the similarities between making this bold economic move and Nixon's recent dramatic announcement that he would be going to Beijing. The China reference truly delighted the president, who loved being seen as a leader who had outfoxed his critics with a big policy coup. Still, when it came to Connally's package of initiatives, Nixon was anxious about both doing everything at once and the timing of an announcement, and he would return to these issues time and again.

On Wednesday, August 4, Nixon, Connally, and Shultz met again in the Oval Office. Connally pitched the entire program again, this time with even more force and enthusiasm. Nixon said the economy was headed in the wrong direction and that they indeed needed to turn the tide. Connally underlined the potential for Nixon to emerge as a great, courageous statesman. While not disagreeing, Shultz raised some questions. Would the wage and price freeze be enforceable? Would it really hold down inflation?

Nixon kept coming back to timing. Shouldn't we wait for Congress to be in session? he asked. The president was also concerned with leaks to the media, his paranoia having been just heightened by the *New York Times*'s June 13 release of the secret Pentagon Papers, a brutally frank and detailed internal history prepared in the Pentagon about America's mistakes in conducting the Vietnam War. Nixon said that all deliberations on this economic package should be confined to the three of them. They all agreed to think more about the plan and its presentation.

SEVERAL DOMESTIC EVENTS heightened the tensions and underlined the need for the president to act decisively. The steel industry had settled a strike with labor contracts that included a whopping 30 percent wage increase over the next three years. The railroad workers had won a 46 percent wage increase over the next forty-two months. Wholesale prices were increasing at an annual rate of 8 percent. On Tuesday, August 3, about a dozen prominent Republican senators—men who would ordinarily have been opposed to wage and price controls—had introduced legislation authorizing them.

More pressure came to bear on the administration on Friday, August 6, when the Joint Economic Committee, chaired by Congressman Henry Reuss (D-WI), issued a report called

"Action Now to Strengthen the U.S. Dollar." It asserted that the dollar was overvalued and that all measures necessary to rectify the balance-of-payments deficits should be considered, with a particular recommendation that wage and price restraints be established. The report said that the IMF should orchestrate a major exchange rate realignment, with countries such as West Germany and Japan revaluing their currencies and the United States devaluing against them, and that Connally should insist that the IMF enforce these actions. "If the Fund [fails] to meet its responsibility, the United States may have no choice but to take unilateral action to go off gold and establish new dollar parities," the report said. As soon as the document was released, Treasury issued a statement that refuted its conclusions, pointing to Connally's pledge made at Munich in late May promising not to embargo gold and not to devalue the dollar. Treasury was being disingenuous, of course, as the Nixon team had by now decided to diverge from Connally's commitments in Munich. Economic historian Allan Meltzer put it this way: "If the market needed to be convinced that the dollar would be devalued, the Joint Economic Committee provided that evidence."

That same day, Connally pushed Nixon again to approve and announce the total package. Nixon warmed to the Texan's enthusiasm. "It's basically psychology," the president said, reassuring himself and signaling his ultimate buy-in to the package. "The country needs a psychological lift. And the psychological lift can only come from doing something." Connally replied, "I agree with that one thousand percent. You have to do something. You have to jerk this country up . . . so that they say we've got a leader here."

Five days later, Nixon met alone with Shultz for ninety minutes in the Old Executive Office Building, where the president had a hideaway office. Shultz talked about Connally's ideas for the dollar and suggested that they needed a strategy to follow the closing of the gold window and the float of the dollar. Shultz was trying to keep the door open for floating the dollar and all other currencies as a permanent solution. In other words, he did not support Connally and Volcker's plan to let the dollar float only as a temporary measure, but ultimately to fix rates again, only with the German mark and the Japanese yen being stronger and the dollar weaker vis-à-vis those currencies. A free-trader, Shultz also argued against the import surtax, saying it would invite retaliation from other countries. At best, he said, the surtax should be temporary. Nixon worried that the concept of devaluation was too complex for the public and that most people wouldn't understand what it actually meant, but that they would certainly warm up to an import surtax. Shultz, staying with the theme of needing to think ahead, said they should have a plan to follow up on the wage and price freeze, which, deep down, he also opposed on ideological and practical grounds.

Nixon worried aloud about getting Arthur Burns's support for the international parts of the package. Burns was, after all, a conservative central banker who, Nixon thought, would favor currency stability and behind-the-scenes multilateral negotiations with other like-minded central bankers, rather than issuing a unilateral set of initiatives and demands, as Connally was proposing. He tasked Shultz with persuading Burns to support the package. The meeting ended with Nixon still wanting to separate the domestic and international parts of the package, doing the domestic part sooner. And he remained wary about devaluing the dollar, a move that could be interpreted as a defeat for the United States.

On Monday, August 9, CEA chairman Paul McCracken sent a memorandum to President Nixon with a comprehensive analysis of the trade and financial situation that described the need for radical policy changes along the very same lines that Nixon, Connally, and Shultz had been discussing. "The current flows of funds to other [financial] centers . . . are merely the thermometer registering the problem rather than the furnace producing it," he wrote. America had lost its preeminent competitive position, he went on to say. Now McCracken mirrored what Connally had been saying. That included allowing the dollar to float—presumably downward against the mark and the yen—and fixing a new parity down the road; implementing a wage and price freeze; and even closing the gold window. McCracken had put aside his fundamental beliefs as an economist about the desirability of floating exchange rates and his opposition to wage and price controls. He knew which way the wind was blowing and wanted to be on the team. It was also becoming clear to everyone that bold action was better than something incremental.

DURING THE FIRST several days of August, the markets had been signaling that a global currency crisis was brewing. Although it was uncertain when it would occur and how it would be implemented, expectations were growing of a dollar devaluation. Belgium, the Netherlands, and France had all asked to convert some of their excess dollars for gold—small amounts, to be sure, but just enough to create anxieties in the market as to what might be next. Dollars continued to flow out of the United States and out of banks abroad holding dollars and into central banks in Western Europe and Japan, where expectations were that the currencies of those countries would emerge stronger versus the greenback. In the August 9–13 period, $3.7 billion moved into foreign central banks. The persistent fear that more central banks would be demanding gold for dollars reached a new pitch.

The *Financial Times* captured the mood: "Markets suggest that we are now at the point of witnessing that most dramatic of all confrontations—when a seemingly immovable object finds itself standing right in the path of a seemingly irresistible force," wrote correspondent C. Gordon Tether. "Thus on the one hand we have the U.S. authorities continuing to insist there is absolutely no possibility of their taking the initiative in devaluing the dollar, its present $35 per ounce link being as immutable as ever. On the other, there is a build-up of factors putting it under pressure to abandon this stand." Still, the markets did not seem to be anticipating anything happening right away. "Should the dollar be devalued," wrote Dan Dorfman in the widely read "Heard on the Street" *Wall Street Journal* column, "the consensus seems to be that it would not occur before next year's national elections [November 1972]."

BY THURSDAY, AUGUST 12, the crisis was at the president's door. During the week of August 9, $4 billion in speculative money fled the United States. On two days alone—Thursday, August 12 and Friday, August 13—foreign central banks absorbed another $1 billion in short-term inflows of dollars. And on the morning of August 12, speculation against the dollar was so strong, and dollar flows into other currencies so large, that the West German mark, still floating from early May, rose to a twenty-year high against the greenback.

Most significantly that day, Great Britain asked the Federal Reserve to insure a portion of its dollar reserves against a change in the value of the dollar. In the crisis atmosphere of the moment, the precise nature of the British request was not clear, but it nevertheless created an air of panic. The Treasury worried that other governments might follow London's request. That would be equivalent to a run on a bank that doesn't have enough reserves. Paul Volcker recalled, "One thing that was clear to me. We were on the brink of a market panic that willy-nilly would force us off gold. If we were going to take the initiative of suspending the convertibility of dollars for gold and present it as a first step of a considered and a constructive package, the decision would not wait."

WHILE ALL THIS was happening, Connally was trying to spend a few days at his ranch outside San Antonio. But at mid-morning on Thursday, August 12, Paul Volcker briefed him by phone, suggesting he return to Washington as soon as possible. Connally agreed immediately. Nixon called him soon after. Connally quickly summarized recent market turmoil, reinforced that the president and his team were in danger of losing the initiative, and said he was returning to Washington that afternoon. Nixon was still clinging to his desire for a piecemeal approach to any plan, but Connally bluntly responded that it would be a mistake. Nixon told him to come straight to the Oval Office when he arrived later that afternoon and that he would talk to Shultz in the meantime.

Connally arrived at the Old Executive Office Building at 5:30 p.m. and joined Nixon and Shultz in the president's hideaway office. Nixon wanted to go over all the options again. He was still talking about a sequencing of the domestic and international parts of the package. He continued to be worried about a negative public reaction to a dollar devaluation. Connally said it was essential to keep an eye on the big picture, not the individual elements. "I think we ought to be primarily concerned about how effectively you convince the American people that you, number one, are aware of your economic problems, number two, that you have thoughtfully considered them not as a piecemeal stopgap emergency measure, but that you have analyzed them in depth, and that you have dealt with them in a substantive manner." Connally made another pitch for a bold approach that contained everything. He said that the problem with doing it piecemeal was that "you will get robbed of a lot of the impact of it." The program will be picked apart, Connally said. People will wonder what's next, as opposed to being stunned and appreciative that the president had his arms around the entire set of challenges. "The biggest thing will be the impact on the American people," Connally continued. "The international thing, hell, we will be in turmoil for a while, as it was before . . . And I don't think you should worry about it." Connally explained that to put it in perspective, exports represented only 4 percent of GDP and that it would be wrong to let the international situation dominate Nixon's thinking. Nixon agreed: "We must not, in order to stabilize the international situation, cut our guts out here." Connally replied, "I couldn't agree more."

Connally argued in favor of moving as quickly as possible, so as to be in control of events.

They turned to the tax on imports, one of the pieces of the Volcker plan about which Shultz himself was conflicted. Connally forcefully made the case for the tariff. "How do you get the devaluation if you don't have something?" he asked, meaning that they needed leverage. "We are dealing with countries around the world who aren't going to let us do what we want to do." They discussed, inconclusively, what legal basis for the import tax they could rely on, but Nixon was nevertheless coming around to seeing its benefits as negotiating leverage. "I feel that we are dealing with some tough countries, like the Japanese, who don't believe we will [take tough action]," he said.

Connally repeated what he had said before: that the plan had something for everyone—business would like the tax credits, the auto industry would like the elimination of the excise tax, labor would like the import surcharge, conservatives would like budget cutbacks, and the average American would like the price freeze.

They talked about Volcker, with Nixon voicing his criticism of Volcker's allegedly purely international perspective and lack of understanding about domestic imperatives. Connally came to Volcker's defense by denying that Volcker thought that way and suggesting that the president speak directly to him.

By now, Nixon finally seemed behind the entire package and an integrated presentation of both domestic and international aspects. He agreed to move as quickly as possible. Preoccupied with what precisely he would announce, he talked about the power of a simple speech. "This isn't something where I think there should be . . . a great big presentation. I think that it's like the China announcement, where action is so powerful, the words should be very brief."

CONNALLY HAD PERSUADED the president on all accounts. Now Nixon said they should have a meeting at Camp David starting tomorrow afternoon. "Now is the time for decision . . . And off we go," he said.

The three men went over the invitation list. It would include the three of them plus others: Burns, McCracken, Peterson, and Volcker; Herb Stein from the CEA, who Nixon thought could be helpful in writing a speech; White House veteran speechwriter William Safire; Shultz's OMB deputies and associate directors Caspar "Cap" Weinberger, Kenneth Dam, and Arnold Weber; plus Haldeman and Ehrlichman. (On Friday night, August 13, they would end up calling Michael Bradfield, legal counsel from Treasury, to come to Camp David early the next morning.)

Everything was ready. Shultz told Nixon that with the Camp David meeting, he had the chance to establish his leadership in domestic and international issues. Later, he told the president, "This is the biggest step in economic policy since the end of World War II." Connally said, "It will be a shot heard round the world, you can be sure of that. It'll be in every town and hamlet." Later, he added, "I'll be the first to admit I'm wrong, but I think this might put your critics so far beyond [*sic*] the eight ball that they are not going to know what to do."

The meeting ended at 7:00 p.m. Twenty minutes later, Nixon called Connally. It was a four-minute call in which Nixon lauded the import surtax and reiterated that he supported the whole package. Connally said that the new program would be like the China initiative: very big and totally unexpected. Nixon told Connally that he should take the lead in the discussions the following afternoon at Camp David.

Haldeman, whom Nixon kept informed all along the way, wrote in his diary on August 13, looking toward the weekend, "We'll cover the [whole gamut of policies] when we do it, so it's going to be quite an earthshattering operation."

Friday, August 13

Friday, August 13, started routinely for President Nixon. From 9:00 a.m. to 10:08 a.m., the president spent time in the Oval Office with an array of advisors, including H. R. Haldeman and Henry Kissinger. He walked into the Cabinet Room to lead a discussion of his National Security Council on the Defense Department budget. Just before noon, he returned to his office for fifteen minutes to pose for photos with a group of White House interns. He engaged in a whirlwind of meetings, each lasting just a few minutes, with, among others, Haldeman; Rose Mary Woods, his personal assistant since the early 1950s; Kissinger again; Attorney General John Mitchell; and George Shultz.

With Shultz, Nixon discussed the meeting plan for Camp David that afternoon and asked whether Treasury had done all the necessary background analysis for the upcoming decisions and follow-up. Haldeman wrote in his diary, "He was concerned about the fuzzy thinking when you get down to the nut cutting, and wondered if the Treasury had the specifics ready to discuss.

He made the point that, in order to know what to do, we have to have a hard analysis that what's being presented will work, and he commented that this is where the intellectuals always fail."

Nixon's concerns were understandable. He was about to make decisions that would shake the foundations of the domestic and international economies: to impose a freeze on all wages and prices in the broad and complex U.S. economy, to sever the price link between the dollar and gold, to encourage a devaluation of the dollar for the first time in post–World War II history, to impose tariffs on all imports in contravention of Washington's twenty-year push for freer trade, to stimulate the economy with a variety of new tax incentives, and to announce offsetting budget cuts. In Nixon's mind, these moves would likely be compared to the achievement of the original Bretton Woods Agreement and to the breathtaking diplomatic about-face he had recently made in announcing his upcoming trip to China. The economic measures he was about to reveal would be all the more shocking because they aimed to go in a direction that was the polar opposite to so many commitments Nixon had made in his first two and a half years in office, especially when it came to wage and price controls and backing the dollar with gold at the $35-per-ounce rate.

Aside from the fact that the policies relating to the dollar and trade were a sharp departure from commitments made by Presidents Eisenhower, Kennedy, and Johnson, they would be enacted without consultation with the United States' closest allies, whose support Washington sorely needed in the ongoing Cold War. In addition, the policy process of arriving at this point had been made under the tight control of Paul Volcker at Treasury, a person and a department Nixon did not wholly trust.

A small group of men had decided what to do immediately following the announcements that Nixon would make, but they had dwelled almost exclusively on how the decisions should be presented to the public. The group had not discussed in any depth what substantive policies would come next in both the domestic and international spheres. "The imposition of the [wage and price] freeze was a jump off the diving board without any clear idea of what lay below," wrote Herbert Stein years later. Referring to the dollar, gold, and monetary reform, George Shultz recalled in 1977, "There was no consensus within our government as to what specific reform objectives should be sought in upcoming negotiations." When it came to the across-the-board import tariff, members of the administration were not in agreement on how long it would last or precisely what concessions from the allies would be necessary to remove it. There had been no careful calculation of the desirable degree of fiscal stimulus that would accompany the other announcements.

To be sure, many big decisions in Washington have traditionally been made in a chaotic way. Leaders are usually besieged by an avalanche of conflicting pressures. On this day, the administration was preoccupied with ending the Vietnam War, critical diplomacy with China and Russia, and potential conflicts between India and Pakistan and between Israel and Egypt. The administration also faced intractable inflation, rising unemployment, a deteriorating balance of payments, and a plethora of complex domestic social programs that Nixon wanted to push. Top policy makers are usually deluged with information from within the bureaucracy and without, and it's an understatement to say they rarely have time to digest effectively what's most precisely relevant to the decision at hand, including all the likely reverberations. Their hand is often forced by events they cannot control at home or abroad. If anything, it would have been highly unusual and almost unprecedented for there to have been an orderly decision process characterized by consideration of trade-offs among different alternative courses of action, negotiating strategy, and analysis of long-term consequences.

Also, Nixon wanted the shock value of a major change in policy, and a full-fledged intergovernmental process that examined all aspects of the economic proposals would never have produced such results. Risk-averse mid-level civil servants would have pressed the administration to follow a more conventional course of action. They could not have preserved the confidentiality essential to preventing the volatile speculation and global panic that would inevitably result from even a hint in the press that the United States was considering closing the gold window.

At some level, Nixon knew all this, having years of experience in Washington. Yet he was still hoping the requisite staff work had been done, that his decisions would be made on solid analytical ground, and that he would have well-thought-out options for next steps. Even if the president had known of the extensive analyses Paul Volcker had overseen, his distrust of the treasury undersecretary would have caused him to look askance at the effort. Nixon had sought Shultz's affirmation on the quality of the staff work because he viewed him as his one advisor who knew how to manage issues and see them through.

THE TRIP TO Camp David that began at 2:29 p.m. is described in the introduction (page 1). Marine One landed just before 3:00 p.m., and the entourage was driven by the Secret Service to Aspen Lodge, the cottage where the president lived and worked when at the retreat.

On arrival, Nixon made a two-minute call to Secretary of State William Rogers, who was in Washington. The day before, one of Rogers's undersecretaries had caught wind of the meeting and contacted Volcker for an invitation. Volcker forwarded the correspondence to Connally, who never replied. The president likely told Rogers that he would be consulting with him later in the weekend on some major announcements he hoped to make, leaving the topics vague.

Inside Aspen Lodge, the chairs were arranged in a large circle for the first meeting, which began at 3:15 p.m. Nixon was dressed in a pale blue sport jacket, while the others wore either sport coats or suits; all wore ties for this first meeting. Volcker stood out in two ways—for his nearly six-foot, seven-inch frame and his white summer suit. Connally looked dapper with a dark plaid sport coat and dark pants. Just before they all sat down, Nixon signed the guest book—an unusual gesture given that he was the host— and he asked everyone else to do the same. He had arranged for his guests to have a Camp David windbreaker jacket with their names and the dates of this weekend stitched on the front. A White House photographer was taking pictures, a rare occurrence at Camp David for a staff meeting. All this, plus the presence of speechwriter William Safire, signaled that something historic was about to occur.

Then Nixon motioned for everyone to sit down, gesturing for Treasury Secretary Connally to be at his right, Fed chairman Burns on his left. As he surveyed the people around him, Nixon would have seen an impressive group: young, smart, experienced, and eager to please. He knew differences of opinion existed among his key advisors. He had worked with them long enough, met with them one-on-one many times, received memoranda from each, and was privy to a constant stream of gossip from Haldeman and Ehrlichman on what each man was thinking, who was working with whom and fighting with whom, and who was suspiciously friendly with reporters and probable leakers.

Nixon believed his biggest challenge would be ensuring that Arthur Burns, with whom he had had a close and some-times tumultuous relationship over the past year, would pledge total support for the broad economic package. Given Burns's personally high standing around the world, not to mention the influence of the Fed's actions on U.S. and global markets, any public dissension by Burns on the package of decisions arrived at that weekend would be disastrous. Nixon had to tread carefully. He knew Burns would be delighted with the decision to impose a wage and price freeze. Still, he worried that Burns would be uncomfortable with suspending gold sales for dollars. Such a move went against the grain of the kind of financial stability to which central bankers dedicated their lives.

George Shultz, too, would be none too happy with the decisions, although he had helped to shape them over the past few weeks. After all, as a disciple of Milton Friedman, Shultz believed in free-floating exchange rates and free trade, and he adamantly opposed anything that smacked of wage and price controls. Nixon wouldn't have worried about Shultz's being on board, however, for the OMB director was a supreme loyalist. Many years later, Shultz recalled that he was satisfied that he had the opportunity to make his case privately to the president and within the government. He also felt that he could have a hand in shaping the ultimate outcomes of the decisions, because so much of the follow-up was still undecided.

Like Shultz, McCracken had been for gradualism and against wage and price controls. But his economic projections were constantly underestimating the severity of inflation and unemployment, and in the previous few weeks, especially in view of currency turbulence, he had grown more willing to follow Nixon's and Connally's push to make some big changes. Besides, McCracken wanted to return to academia. He had always seen himself as a professor as much as a public servant. Still, it was unlikely he would want to leave Washington falling out of favor with the administration.

Except for Connally, Peterson was most focused on the public presentation of the new initiatives. He was, after all, a marketing genius, having honed his skills as CEO of Bell and Howell. On such issues, Nixon wanted Peterson's input and guaranteed him an active role in the deliberations. But Peterson wanted to play a bigger role than just chief marketer and was hoping to present broader and longer-term issues. Unfortunately for him, Nixon had already begun to tire of what the president considered Peterson's long-winded analyses that didn't respond to Nixon's preoccupation with policies that could produce visible results before the next election.

Nixon felt that Volcker typified a Treasury that was risk averse and unimaginative and that given his druthers, Volcker would oppose most of the decisions that were about to be made. The president resented Volcker's intense international perspective. He had no time for Volcker's view that many of the issues surrounding the dollar were the result of lax policies at home and that higher interest rates and a smaller budget deficit were the medicine the United States and the world needed. And Nixon was convinced that Volcker would fight for the status quo—especially when it came to the dollar and its link to gold. Still, the president also knew that Connally held Volcker in exceptionally high regard, and Nixon counted on Connally to keep his undersecretary in line.

On the subject of a wage and price freeze, Nixon himself had been adamantly opposed to such policies. Speechwriter William Safire recalled in his book *Before the Fall*, "In every economic speech I had ever worked on with him, there was a boilerplate paragraph on the horrors of wage and price controls, how they would lead to rationing and black markets and a stultifying government domination of the economy." However, Safire also said that Nixon told his staff on the eve of this meeting that "Circumstances change. In this discussion, nobody is bound by past positions."

Nixon expected Connally to be the battering ram at the meeting, if one were necessary. When it came to the across-the-board tariff, Nixon knew that Connally was enthusiastic, but no one else would likely be; still, he felt his treasury secretary would carry the day. The president understood that the Texan was without any ideology. He was a pure politician, focusing on what would work and not caring whether he was departing from past policies. After all, in late May, Connally promised the world at a conference in Munich that there would be no dollar devaluation. A month later, following a Camp David meeting on economic policy, he said there would be no wage and price controls. He had now abandoned both these positions without any hesitation.

Congress was a factor, too. Nixon faced a Democratic majority in each chamber, and under other circumstances, he might have expected opposition to his major initiatives. But, almost everything Nixon was proposing was plucked from Democratic thinking. Congress had provided authority for wage and price controls a year ago. While, at the time, Nixon said he didn't want such authority, Connally thought it might come in useful—and now it had. On exchange rates, Henry Reuss (D-WI), who headed up the Joint Economic Committee and who was among the most active and knowledgeable members in Congress on the dollar, had been urging more flexibility in exchange rate arrangements and, indeed, a lower value for the dollar. When it came to the import surcharge, the Democrats in Congress had turned sufficiently protectionist that this measure was sure to please them. As for fiscal incentives to stimulate the economy, Nixon knew that Congress would surely go along; it was more interested in growth and job creation than in fiscal rectitude.

HIS FEET RESTING on a small footstool, Nixon began the meeting by asking Paul Volcker to bring everyone up to date on the currency markets and the gold situation. It is likely that

Volcker had in his hand a Treasury report prepared that weekend. The key trends were a steady deterioration in the merchandise trade balance, leading that year to the first trade deficit since 1893 in comparison to a postwar record trade surplus in 1964 of $6.8 billion. The document showed that the U.S. stock of gold had deteriorated to almost $10 billion from a high of nearly $22 billion in the late 1950s. This $10 billion was de facto collateral for some $40 billion held by foreign central banks and eligible for conversion into U.S. gold. No one thought that all foreign governments would rush through the door to demand this exchange. In fact, in March 1967, West Germany had pledged not to do so in deference to the U.S. defense commitment, and it was a near certainty that Japan would abstain for the same reason. Yet with the deteriorating U.S. economic picture, the administration's concerns did not go away. Everyone knew that in such tense financial conditions, a monetary firestorm needed but a match to ignite it. And who could be sure that, under their own domestic pressures, the West German and Japanese governments would not change their minds?

Volcker told the group a harrowing story of an event that occurred the day before. He and Connally had received word from the New York Federal Reserve (the part of the Federal Reserve System that handles international transactions) that Great Britain had just asked for "cover" for $3 billion of its dollar reserves. What did "cover" mean? No one was certain, but there were two main possibilities. Either London wanted a guarantee that if the dollar were devalued, the United States would reimburse Great Britain for the amount it lost on the dollar reserves it held. If, for example, Great Britain were holding $3 billion of reserves in greenbacks, and if the United States devalued against gold by 10 percent, then the value of Britain's reserves, measured in gold, would decrease by $300 million, and Uncle Sam would owe Britain that amount. Or, "cover" could be interpreted as London's wanting $3 billion worth of gold now. The precise request wasn't clear to either Volcker or Connally, but in this meeting, the specter was that an attack on gold could well happen, with Britain, the United States' closest ally, coming first and others following. It would be nothing less than a run on the bank. In fact, it turned out that Britain's request was neither of these, and the communication to Volcker had somehow been garbled or misinterpreted. The Fed was being asked for something far less dire and far more easily accommodated: a guarantee of protecting $745,000 of British reserves against a downward change in the value of the dollar. But with that message not clearly understood immediately, and because of Volcker's and Connally's misinterpretation, the tension and sense of urgency at Camp David were seriously heightened.

AFTER VOLCKER FINISHED his overview, Nixon took over. He began boring down on the issue of secrecy. The only telephone calls allowed at Camp David that weekend were for the purpose of getting vital information pertaining to the development of policies discussed there. Nixon demanded radio silence until he made his announcement, which he was envisioning doing on Monday night. "He referred to China," Haldeman wrote in his diary, "and said we had responsible people, and highly intelligent people, and not naïve people dealing with the China thing, and consequently it didn't leak."

Nixon rambled for fifteen minutes. His key point was that the international economic situation had forced Washington to act to address the underlying causes of their problems, such as domestic inflation, fiscal deficits, and soaring wages and prices. He emphasized that they had to deal with an import tax that would have to go to Congress. Haldeman wrote, "He said it would not be relevant to know how we got here, that analyzing the actions would do no good, instead we have to find solutions." Then Nixon asked Connally to outline the decisions to be considered.

It is clear what must be done, the treasury secretary said. There was no scene setting; he assumed that everyone present already had the complete background for all the issues. Yet Connally's shorthand rendition could also have been interpreted as his expecting everyone to quickly sign off on it.

Connally proceeded: They had to close the gold window but not change the price of gold. Everyone knew what this meant. The United States was not going to devalue the dollar in a technical sense by changing the $35-an-ounce relationship, but Washington would demand that other countries revalue their currencies, creating the effect of a dollar devaluation. Thus, if, at the time of the Camp David meeting, the dollar were equal to 360 Japanese yen, and if Japan were to revalue by 20 percent, the new exchange rate would be 288 yen to a dollar (360 minus 20 percent of 360). In other words, there would be fewer yen to the dollar, making each yen worth more dollars than had been the case before yen revaluation. What would all this mean for trade? One dollar would buy fewer Japanese goods than it used to, slowing imports into the United States. At the same time, it would take fewer yen to buy a dollar's worth of U.S. goods, making U.S. exports cheaper for the Japanese and thus causing Japanese consumers to buy more goods from the United States. By pressing other governments to revalue their currencies, and by assuming the dollar would remain steady relative to gold, the United States would be spared the political embarrassment of diminishing the value and status of the dollar. It would also avoid the need for Nixon to persuade Congress to change the dollar price of gold, something that, under U.S. law, only it could do.

Connally then enumerated a series of fiscal measures that would stimulate the economy, some added to the list he had compiled the day before. He talked about reinstating an investment tax credit of 8–10 percent. He said the United States would repeal an excise tax on autos. It would institute a tax-incentive program for exporters and loosen antitrust constraints on them abroad. And last, it would remove a host of controls on outgoing capital flows that had been established by the two previous administrations. These controls were enacted to make the balance-of-payments deficits smaller than they otherwise would have been. Connally saved the wage and price freeze for last, asserting that it would be a key component of the package. Summing up, he said, "Such a program will leave a clear impression that this [plan] has been analyzed in depth, [and has] not [been] just a reaction to pressure." Also, "It would be an act of great awareness, great statesmanship and great courage, and must be presented to the people that way."

Nixon jumped back in. He emphasized that the tax breaks would have to be matched with budget cuts; he was adamant about not increasing the deficit. He turned to the wage and price freeze, saying it had to be temporary and that they needed to think through what would come next, even if they did not announce it now. He then turned to the import tax: should it be applied selectively, or to everyone? He also asked what the underlying legal authority was and the role of Congress.

By this early point, while a substantive discussion of the import tariff and provisions relating to gold and the dollar ensued, it had become clear that the big questions had already been decided in Nixon's mind. Perhaps the full picture was already obvious, but ordinarily when such complex decisions were made, someone set the context in order to channel everyone's thinking at that moment. That kind of orientation didn't happen here.

Nixon and Connally were forging right ahead. Perhaps the others did not need to be reminded that this meeting marked a clear inflection point in postwar American history, a moment in which a presidential administration took concrete decisions to lessen the enormous burdens it had assumed since the end of World War II. After all, Nixon had already announced and widely promoted the Nixon Doctrine of reducing U.S. defense commitments to defend all of the free world and how a similar retrenchment must happen in the economic sphere. He had carefully read and widely circulated Peterson's report "The United States in the Changing World Economy," which had clearly documented the new, changing, and hypercompetitive global market the United States now faced. It's a good bet that the men sitting in Aspen Lodge that afternoon also fully understood the interlocking nature of the domestic and international decisions they were going to make—the link between inflation and the dollar, the relationship between wage and price controls and the import tariff. All had a sense that any action was better than no action and that the United States would have to take the chances that came with a preemptive stance. Wrote *Time Life*'s Hugh Sidey later that month, after having interviewed many of the participants, "The men around [Nixon] were to be the tacticians in a campaign already conceived in its broader outlines."

THE DEBATE OVER the import tax took place on two levels. Was it advisable in the first place? Volcker strongly opposed it. He thought it overkill and that Washington should hold it in abeyance to be used later if additional leverage beyond the closing of the gold window was needed to persuade others to revalue their currencies. McCracken also seemed lukewarm about the tariff.

Shultz didn't speak on this point. He had opposed the tariff on ideological and policy grounds in previous meetings with Nixon, but the day before, he flipped his position and said it was okay, as long as it was just a temporary negotiating weapon. Connally forcefully supported the tariff, believing a full package now would elicit a better public reaction. Nixon said that if the administration didn't impose it, Congress itself would, and it would be better to have control over its form. Burns supported Connally and Nixon as long as the import tax was temporary—say, six months; otherwise, it would elicit too much foreign hostility, he said. Connally and Nixon agreed, but Connally didn't want to decide now on how long such a tariff would remain in place.

When it came to trade, Nixon was obsessed with textiles, given his 1968 election commitment to southern senators that he would curtail imports. He asked Shultz to address the textiles issue, thinking that in the measures he would now take, textiles could be solved as well. Shultz explained that, under various U.S. trade acts, Nixon didn't have legal authority to curtail textiles—or to tax other imports, for that matter. That would require new legislation. The possibility was raised that he could use authority under the 1917 Trading with the Enemy Act, which was designed for national security and under which the president had blanket permission to do almost anything in the trade arena. Nixon didn't like the connotation of the words "trading with the enemy" as applied to an ally like Japan. "No," he said. "That smacks wrong from the point of view of international leadership . . . My long-range goal is not to erect a 10 percent barrier around the U.S.—that would be retrogressive—but to set a procedure that lets us go up and down with room for negotiation . . . If we could [implement a surtax on imports] and at the same time close the gold window, then we would provide the basis of negotiation. It gives you more stroke" (meaning more leverage).

Peterson raised an alternative under international trading rules sanctioned by the General Agreement on Tariffs and Trade (GATT). Under GATT, all the United States had to do was declare a balance-of-payments emergency as a justification for impeding imports on a temporary basis. In that event, by agreement among the parties to GATT, other countries could not retaliate. Nixon thought that was a much better way to go, and Burns agreed.

NIXON SAID IT was time to discuss the gold question. McCracken argued that if the gold window were closed, the market would force the dollar to depreciate against other currencies—a good thing given the balance-of-payments deficits. Yet, the imposition of an import tax would tend to cause fewer U.S. purchases and fewer dollars to be paid to foreigners, thereby decreasing the supply of greenbacks abroad and possibly raising the value of the dollar. Connally said he understood this inconsistency, but that the United States should close the gold window anyway because they needed a club to hang over the heads of other governments. Connally was implying that negotiating leverage came with the idea that, sometime in the future, after its balance-of-payments goals were achieved, the United States would restore, at some price ratio, a dollar–gold link. This was everyone's unspoken assumption at that point, although it was never articulated.

Burns intervened: "At the right moment, I want to question the judgment of closing the gold window." The president replied, "Let's talk about it right now." Burns replied, "No, let's finish with the border tax first."

Shultz said he would calculate the revenue that an import tax would bring in and also find a connection to the Japanese textile negotiations.

Nixon turned to Burns on the gold question. "Arthur, your view, as I understand it, is why is it not possible to do all of the things that are at the heart of the problem and then to go to close the gold window [later] if needed? The Treasury objection to that is reserve assets would be depleted quickly."

Burns embarked on a fifteen-minute lecture about the dollar. He admitted that Connally and Volcker could be right about closing the gold window, but that he couldn't agree. Embargoing gold sales now was unnecessary because the steps that Nixon was already taking were powerful enough—the wage and price freeze, import tariff, tax cuts, spending cuts, and control of the budget deficit. These steps alone would electrify the world, Burns argued. After hearing about all the other initiatives the administration was taking, the flight from the dollar would cease, and no one would want to cash in their dollars for gold. Burns was no doubt thinking about other central bankers and their interest in not rocking the boat.

Turning to the risks of closing the window, Burns expressed fear of the unknown reaction in the markets and among U.S. trading partners. No one could tell what would happen to the stock market. No one could know if foreign countries would resort to trade retaliation. "We release forces we don't need to release," Burns said. He suggested that Nixon would take a big political hit, and that the Russians would cheer the move as a sign of the collapse of capitalism. He suggested that Paul Volcker be sent on a mission to negotiate a realignment of exchange rates, rather than the United States defiantly embargo gold.

Nixon intervened. "In reading over the years on this subject, I have never seen so many intelligent experts who disagree 180 degrees. George [Shultz] and others like the floating idea (which would make gold irrelevant). Arthur says to get in a good negotiating position and then deal"—by this, Nixon meant negotiate to reinstate the present fixed-rate regime, albeit with an exchange rate realignment that would make the dollar relatively weaker than it had been and, hence, create more competitively priced American exports. Nixon said he was worried that unless the gold window were closed, speculators would attack the dollar by demanding what little gold the United States had left.

Connally interrupted. "What's our immediate problem? We are meeting here because we are in trouble overseas. The British came in today to ask to cover $3 billion, all their dollar reserves. Anybody can topple us—anytime they want—we have left ourselves completely exposed."

Connally argued for the United States to pursue its own interests as decisively as it could. Volcker said he agreed with Burns but thought that the international impact of closing the gold window could be managed if the United States followed with a genuine effort to negotiate an exchange rate realignment. "I hate to do this, to close the window," Volcker said. "All my life I had defended exchange rates, but I think it is needed. But don't let's close the window and sit—let's get other governments to negotiate new rates." Volcker was expressing a deep-seated concern. He was afraid the United States would suspend gold sales and that all currencies would then become untethered, floating against one another with no subsequent plan for realignment with a new set of fixed rates.

Connally turned to Burns. "Why do we have to be reasonable?" he asked.

Burns: They can retaliate.

Connally: Let 'em. What can they do?

Burns: They're powerful. They're proud, just like we are.

In all the discussion, no one asked whether closing the gold window was a permanent move or whether it was just a negotiating lever.

The discussion was fluid and energetic. Everyone wanted to be involved, as it meant relevancy to the most significant set of economic issues they might ever be party to in their professional lives. At one point, according to a description of the meeting by Hugh Sidey, the *Time Life* columnist who covered Nixon, "Mc-Cracken was taking so many notes that he ran out of tablet paper and started to write on the back of envelopes."

Herb Stein brought another issue into focus: the potential reaction of the labor movement. This was an opening for Nixon to opine about the overall response of the American people to what the Nixon team would be proposing. The president said that labor would applaud the border tax, the price freeze, and the tax incentives to get the economy moving.

Nixon returned to the gold window. He acknowledged that it was complicated to explain this issue to ordinary people and said the media would jump on it to frighten people. Thus, it was unknown how that part of the deal would ultimately be received by the American public.

Burns said that while academic economists favored floating exchange rates, where gold would play no part, business economists were on balance opposed. They liked stability and predictability.

Burns then went back to the overall foreign reaction to what Nixon would propose: Other countries could hurt us with their retaliation, he said.

Connally replied that they were already hurting us.

PETERSON CHANGED THE subject, sensing that the main course of action had been resolved and that what Nixon cared most about now was how the new package could be presented with maximal political effect. Peterson said the speech should contain the theme of sacrifice and statesmanship. Burns chimed in: what about the idea of looking long term to create a set of arrangements that would overcome the constant monetary crises the world had experienced? Nixon wasn't interested in either thematic suggestion. Such subjects should be in a background briefing given to the press and not in the speech, he said. He was focused on the more immediate issues and how to present them so American citizens believed they themselves would benefit right now.

Then Nixon held forth on the speech he wanted to make. Peterson was correct: the public presentation of the proposals was what was most on the president's mind. When you have a lot to say, Nixon explained, you lose something if you say a lot. A ten-minute speech would do it—"crisp, strong, confident."

Peterson suggested that Nixon focus on American competitiveness, an agenda the former had been pushing for over the last several months. He urged the president to talk about the development of new American technology. But no one seemed to pick this up, and the conversation reverted again to the gold window.

NIXON SAID THAT many businessmen would understand and applaud stopping the exchange of gold for dollars, but he reiterated that the media would scare people by talking about devaluation. Burns said that the real test would be in the markets. If the price of gold shot up in the free market, then everyone would say the dollar was being devalued, regardless of what the media did or didn't say. Volcker said they had to be prepared. The United States should make it clear that gold was not so critical anymore to currency management. He didn't go into detail, but he was surely thinking that the new SDRs (assets issued by the IMF) would be a long-term substitute for gold. Volcker raised the issue of whether the United States should sell some gold if the price shot up, thereby demonstrating that gold was no longer so central to Washington.

McCracken jumped in. There was no doubt that the general reaction to closing the gold window would be negative, he said, because most people wouldn't understand it. Nevertheless, he thought that the positive reception to a wage and price freeze would counterbalance the gold issue. Nixon asked Peterson what he thought about the public reaction to closing the gold window. "They will worry," Peterson replied. Nixon repeated that the media would attack them on the gold issue.

Connally intervened. He said that the discussion so far implied that the United States had viable alternatives, which was not the case. We have to deal with the situation we have, he said. That's why we are here. We need to move now, not a day later. "We don't have a chance unless we do it," he said. "Our assets are going out in a bushel basket." We should close the gold window now along with announcing the wage and price freeze, Connally continued. Otherwise, we are at the mercy of the moneychangers, the central bankers.

Burns seemed to take offense and used the occasion to protest once again any closing of the gold window. He explained that the moneychangers were not all bad. In fact, he said, he knew them well and got reports from them regularly. They have said all along that if the United States took bold steps to bring inflation under control and managed its deficits, they would be happy and would not try to exchange their dollar reserves for gold. He reiterated that, given everything else the president was to announce, closing the gold window just wasn't necessary right now.

Connally replied that the moneychangers might not assault us on gold all at once, but "they'll nibble us to death." Peterson chimed in to support Connally. He said it was a bold package in its entirety, including the closing of the gold window, and that they'd look foolish leaving out that provision now if they had to resort to it later.

NIXON STEERED THE conversation to budget matters. He reeled off a number of spending cuts he wanted to make, believing the appearance of fiscal austerity was key to the overall package. He started by indicating which of his high-profile domestic programs would have to be deferred, including his administration's attempt to modernize welfare programs and his proposal for a "New Federalism"—the idea that the federal government would return money to the states and make good on Nixon's commitment to decentralize government decisions. He also wanted to defer the next federal pay raise and institute a freeze on federal hiring and a 5 percent cut by attrition. Haldeman's diary entry puts it this way: "Then [the president] issued an immortal quote. 'The average person thinks the federal employee is unemployed anyway, so it won't hurt unemployment.'" Nixon charged George Shultz and his deputy Cap Weinberger to work out all these cuts. Shultz said they could cut between $5 billion and $7 billion. Nixon proceeded to discuss the cuts in great detail, displaying his deep understanding of the arithmetic of the federal budget and underlining his commitment to demonstrating he cared about reducing the deficit.

THE DISCUSSION VEERED back to the wage and price freeze. Nixon asked Shultz to start off and asked Burns—who had been so vocal on the need for controls over the last year, much to Nixon's annoyance—to chime in. The OMB director said the freeze would work best if it were related to the international crisis, particularly to the inflationary impact of a cheaper dollar and to a surcharge on imports. The freeze should be short, simple, with no exceptions, and they would need to figure out what came next. Nixon stated his concern with not creating an unwieldy bureaucracy that would put the economy in a straitjacket. Shultz agreed, suggesting that the freeze and whatever program followed should be overseen by a tripartite board of business, government, and labor, with no more than seven members. Burns concurred, adding that Shultz's ideas were remarkably similar to his own.

Many issues emerged concerning the length of the freeze, whether its duration should be announced, and what legal authorities the president had. It was clear the freeze would cover wages and prices, but what about dividends (which would accrue to well-off investors) and interest payments (which would accrue both to investors and to those who saved money, possibly for retirement)? It wasn't clear that the president had legal authority to control dividends, so everyone agreed that companies would be asked to abstain voluntarily from issuing them during the freeze. The control of interest rates was not belabored, perhaps because the participants thought it would be too difficult to get legal authority to do that.

BY NOW IT was around 6:00 p.m., and the group had been talking without a break for almost three hours. They had focused on most of the components of the package. Despite the visceral dislike of wage and price controls on the part of many of the men in the room—Nixon, Shultz, and McCracken among them—no one opposed them, although considerable concern was expressed over their implementation, about which they and their colleagues had given little thought. Fiscal investment incentives didn't get much attention at all, and Nixon himself seemed to be the strongest proponent of budget cuts. The import surtax had proved just mildly controversial to the group, even though it constituted a fundamental change in U.S. trade policy. Overall, closing the gold window was still the biggest area of disagreement.

When it came to next steps, Volcker again pleaded that the period of closing the gold window and the implementation be used to negotiate a more permanent change in the international monetary framework. Shultz said the same about using the wage and price freeze to figure out the follow-up machinery.

Nixon returned to the speech he would give. He ruminated on the idea that the United States could have prosperity without war, that it should operate at full capacity even as the Vietnam War came to an end. He wanted to begin the speech with the question of what needed to be done to achieve peace and prosperity. What were we up against? Haldeman recorded these thoughts as expressed by Nixon: "It's a very competitive world. Is the United States going to be a great nation, number one? Don't assume that unless we are ready to meet the challenge." Nixon then talked about the things that America needed to do—deal with unemployment, deal with inflation, get the economy growing, etc. "We're not going to allow the dollar to be destroyed," Nixon said, and he then wove in the role of a wage and price freeze and the other steps he would take. "We say, here's a problem, we're going to deal with it, [we will] show we're willing to do things differently." Safire wrote, "The American people must give their moral support to the effort. What can you do?" Nixon said as if he were addressing the television audience. "Labor, support the freeze. Businessmen, invest. Consumers, buy. All of us get off our butts."

The president then said that William Safire would edit the speech, and that he should work with Herbert Stein and Peter Peterson. Nixon reemphasized that the speech should be very short—say, fifteen hundred words—for maximum impact. Burns suggested that the speech be moved up to Sunday night, before markets opened. Nixon worried that might make it appear that the administration was panicking. The timing issue was left open.

They talked about who would brief the press. Connally and Shultz were to be out front, to begin with. All others would play big roles. Nixon said that Secretary of State Rogers needed to be informed and that a few calls should be made to Congress. Business and labor would also need to be contacted. "When you background on this," said Nixon, "put 75 percent of your effort on TV . . . I have not had anybody working for me in the press area who understood the power of TV."

Shultz suggested that they organize themselves into working groups to refine the details of the proposals. Connally and Volcker were to examine monetary, trade, and tax policies, McCracken and Stein would look at the wage and price freeze, and Shultz and the three associates would deal with budget issues.

THE MEETING ADJOURNED at 7:00 p.m., at which time Nixon, Burns, Connally, Shultz, McCracken, and Volcker remained to allow Burns to talk more about his opposition to closing the gold window. The others walked over to Laurel Lodge for dinner.

The president was going out of his way to ensure that Burns had as much time as he needed to present his views. The Fed chairman went through his arguments again about why they should hold back now on closing the gold window. He expressed concern about another issue, too. Connally had proposed abolishing all direct controls on foreign investment and lending that had emerged in the Kennedy and Johnson administrations to curtail dollar outflows. Burns thought foreigners would see this measure as an aggressive effort to encourage the outflow of dollars to drive down the value of the U.S. currency. Moreover, he said, labor would see the release of greenbacks as a way for companies to invest abroad and outsource American jobs. The president agreed with him and dropped the idea. Removing capital controls would not be part of the new program.

Burns felt good after the meeting. He wrote in his diary that the president had been very deferential to him. Nixon kept referring to their long friendship and "wistfully commented that the chair in which I was sitting was the one occupied by Khrushchev during Camp David discussions with Eisenhower during the mid-50s." Burns continued: "I assured the President that I would support his new program fully. I could do this readily except for gold suspension; and, although the President was unaware of this, I could not question what he did on gold—publicly." The implication seemed to be that the Fed was not in a political position to have a confrontation with the administration over the currency question, because the resulting chaos in markets could lead to a global financial meltdown. Burns went on to write, "Moreover, I was aware of the margins of doubt on this question, and I could not be sure that my position was right and his wrong." When characterizing Burns's agreement to support Nixon, historian Allen Matusow described it this way: "With that pledge, Burns gave Nixon what he wanted from Camp David above all. Burns was back on the team."

BURNS AND THE others then walked over to Laurel Lodge to join the rest of the group at dinner. Nixon remained in Aspen Lodge and ate with his personal secretary, Rose Mary Woods.

William Safire was anxiously awaiting the outcome of the private meeting with Burns about whether to embargo all sales of gold for dollars so he could proceed with the speech. As soon as he arrived at Laurel, Volcker cornered Safire and said that he should write the speech as if the gold window would be closed.

As dinner proceeded at Laurel, Nixon called Haldeman to say that he wanted to give his speech on Sunday night and not Monday. Contrary to what Volcker had just told Safire, Nixon said that he was leaning toward Burns's original position not to close the gold window at this time and to wait to see if the other measures alone were enough to keep foreign governments from exchanging their dollars for gold. Nixon knew that Connally would be opposed to this postponement, and he asked Haldeman to see if Shultz and Ehrlichman could persuade the treasury secretary to go along.

Some levity and light banter relieved the tension at dinner. Volcker said that Nixon's prohibition against their making any phone calls was a good thing because speculators could make a fortune if they knew what was going on there. Haldeman leaned over and jokingly asked, "Exactly how?" Volcker asked Shultz how big the budget deficit was. Shultz replied that it was about $23 billion and asked why Volcker wanted to know. Volcker said that if Shultz gave him $1 billion on Monday and a free hand, he, Volcker, could make up that deficit by speculating in the money markets.

At dinner, the conversation turned to the press conference that would have to occur after the president's speech. Safire started to pose some potential press questions, and Connally began to answer them. Occasionally, others stepped in with answers of their own. Some of Safire's questions were intentionally snide, the speechwriter recalled, saying, "I began to thoroughly enjoy flaying the Secretary of the Treasury."

During the dinner, Ehrlichman and then Safire were alternately summoned to another room. In each case, Nixon was calling to find out what was going on. From Ehrlichman, the president heard about the mock press conference, and when he spoke to Safire, he complimented him for starting the questioning, telling him, "Ask all the tough [questions]." He then proceeded to give detailed guidance on what Safire should ask and who should be prepared to answer him. "Tell McCracken and Shultz to downplay rent control. We're doing it, but I don't want to screw up the housing boom—too many jobs depend on it," Nixon said. Then he asked Safire, "You're sitting next to Arthur? You know, Arthur is a fine man. Really a fine man. Tell him I said that."

After the meal, Connally stood up and clinked a glass to end the conversation. He asked everyone to break up into working groups in separate meeting rooms at Laurel.

Sometime that evening, Volcker made a phone call to Michael Bradfield, a deputy legal counsel in Treasury specializing in international finance and trade, to come up to Camp David the next day. Volcker emphasized the secrecy of the meeting. Legal expertise would be needed on several fronts, particularly on the authority to impose a tax on imports, and Bradfield, who had worked with the Volcker Group, had already done a lot of that spadework.

As the working groups met, Safire walked back to his cabin to work on Nixon's speech, writing most of the night in order to have something ready for the president by breakfast.

At 9:44 p.m., Nixon called Haldeman again and said he'd be available after 8:30 a.m. the next day to meet with the group. Haldeman recalled, "Nobody had any intention of meeting with him further tonight, but he seemed to think they might, so he was turning it off."

IT HAD BEEN a long day for everyone. They had come to Camp David to change not just the direction but the structure of the U.S. and world economies. It had been a particularly good day for Nixon. With Burns now seemingly in his corner, the president had every reason to believe his team was now aligned behind him. Given their disparate views on the international monetary possibilities—fixed versus floating rates? some role for gold or none?—given the global shock that would accompany the imposition of a surtax on imports from the world's heretofore champion of free trade; and given the strong antipathy most had for wage and price controls—given all that, getting everyone pointed in the same direction was no small feat.

Saturday, August 14

Richard Nixon awoke at 3:15 a.m. and walked into his den to work on his speech. It was still cool outside, somewhere in the 50s, with projections that temperatures would get to the high 70s or low 80s. As was his habit, the president began organizing his thoughts by writing an outline on a long yellow legal pad. He filled both sides of three sheets with notes and then proceeded to dictate his thoughts into a Dictaphone.

At 4:30 a.m., he called Haldeman, apologizing for waking him up. In a conversation that lasted about ten minutes, Nixon told his chief of staff that he was outlining the speech and knew exactly what he wanted to say. He said he needed to see Safire first thing in the morning, and until then, he didn't want his speechwriter to lock into any themes of his own.

Nixon told Haldeman that he'd thought about what Arthur Burns had said about the gold window and that he had finally made up his mind to close it—reversing what he had told Haldeman the night before. And he would give his speech on Sunday night.

Then the president read Haldeman some of his notes. "I've addressed the nation several times regarding ending the war. Because of the progress we've made, this Sunday evening it's an appropriate time to address the nation regarding peace. America today has exciting prospects, a full generation of peace, and a new prosperity without war." He continued with a few other lines.

At 5:45 a.m., Nixon put on a robe and walked toward Witch Hazel, the cabin where Rose Mary Woods was staying, intending to slip his Dictaphone tapes under her door. On the way, he saw a navy chief stealing a swim in his pool. Nixon said good morning, and the startled chief answered, "Yes, ma'am." The amused president asked the sailor to take the tapes to Witch Hazel and put them inside the screen door, rather than do it himself. Then Nixon went for a half-hour swim.

Rose Mary Woods woke up early and found the Dictaphone tapes. When Safire arrived at her cabin at 7:00 a.m., having himself worked on the speech most of the night, he was startled to find her typing away. When she finished, she handed him the president's memo containing precise instructions to Safire about the speech he wanted to give.

At 8:40 a.m., Nixon called Haldeman again, wanting to be sure that Safire knew that the form and structure of his notes should be scrupulously followed. The president liked his own gutsy rhetoric and wanted to keep that feeling. He still wanted Safire to put more zip into his speech, but not to go too far. "He had a great line for Safire," Haldeman wrote: "'Don't make it brutal and beautiful, rather, brutal and effective.'" Nixon said that Safire was to show the whole speech to no one, except to check technical points, and then with only the person most knowledgeable about that section.

Nixon also brought up with Haldeman the issue of the dollar–gold link. The president was still anxious about how closing the gold window would play with the public. Haldeman recorded, "[The president] said the major problem this morning was the gold float, and he's almost around to Connally's view that we should take the risk and take the heat [of closing the gold window]." He further wrote that Nixon then came up with a phrase that we need to "defend the dollar against the international speculators." Nixon said he was worried that Volcker and Burns would want to defend those same speculators. He continued to feel that both were too concerned with the international dimension of policy, rather than with his domestic political priorities.

Nixon gave Haldeman another rendition of the themes he wanted to pursue in the speech. He said the theme of a "new prosperity" would have three elements: jobs, stop the rising cost of living, defend the dollar against speculators. The import tax would come in the section on speculators.

He again conveyed instructions for Haldeman to give to Safire. He felt now that twenty minutes was the maximum length, twenty-three hundred words. He wanted Safire to add facts and figures to bolster Nixon's ideas and not much more.

The president then abruptly changed the subject to consultations with outsiders and cabinet members. He relayed the following directions to Haldeman: Ehrlichman was to call New York governor Nelson Rockefeller, a prominent Republican. Nixon would be proposing budget cuts to offset some of his new stimulus proposals, and he wanted to allay the governor's fears about postponing some federal assistance Nixon had promised him. Tell Rockefeller that it's just a "tactical matter," Nixon said. He asked Haldeman to call Secretary of State Rogers and "tell him that we are going to be taking some strong action, but conciliatory, for the purpose of building a new monetary system, that we want competition, but we want it to be fair." He also wanted to know whether Rogers would agree to a 10 percent cut in foreign aid as part of the budget-cutting package. Nixon talked further about whom should be consulted in advance by phone calls and special briefings.

The president turned back to his concerns about Safire. He wanted the speechwriter to insert a line in the speech about trade and how the balance of payments had been eroded for fifteen years because of the competition of other nations. Emphasize "the exchange rate discrimination against the U.S.," Nixon said.

AT THIS POINT, Nixon was riveted on the speech. It would be his own product—the substance, the structure, the tone. Still, it was William Safire who would shepherd it through and who would labor over many of the key phrases and check its contents with key advisors. Safire would be the central interlocutor with Nixon all the way through the writing process.

The two men went way back together. They first met in 1959, at an American exhibition in Moscow designed to showcase the wonders of the U.S. economic system. Nixon was Eisenhower's vice president, and Safire was a press agent for a company called All State Properties, an American homebuilder. Safire witnessed Nixon debating Soviet premier Nikita Khrushchev in the nearby RCA exhibit, which had been equipped with TV cameras. The Russian leader was berating Nixon about the shortcomings of capitalism. Nixon, the consummate street fighter, was inexplicably behaving like a restrained diplomat. He immediately realized he would be seen by audiences around the world as having lost a debate with the Soviet leader. "Nixon came out of the TV studio sweating profusely, knowing he had 'lost' [the debate] and anxious to make a comeback," recalled Safire years later. Both Nixon and he knew the vice president needed to recoup fast from his humiliation.

Seeing an opportunity to promote his showcase house and also help the vice president, Safire guided Nixon to his company's homebuilder pavilion, with Khrushchev trailing behind. In the kitchen, Nixon tore into Khrushchev in what became known as the "Kitchen Debate." Reporter Harrison Salisbury of the *New York Times*, the one pool reporter, caught the mood and helped Nixon's powerful performance make headlines around the world. A photographer took a picture of Nixon jabbing his finger into the Soviet leader's chest. Thanks to Safire's assistance, the vice president was seen around the globe as defending American values.

Safire became an informal advisor in the years between the Kitchen Debate and Nixon's election, after which he was hired as a speechwriter. Nixon was the first president to build a large team of speechwriters, one replete with researchers, and Safire became a mainstay of the team. He was often picked for economic messages, but wrote about many other topics, including federalism and violence in America.

Safire was asked to sit in on many of Nixon's meetings, and to attend banquets and other social events to try to capture Nixon's lighter and more informal moments for anecdotes that could be used to humanize the president. At a time when the administration saw itself as at war with the press, Safire argued that Nixon should maintain open channels to news outlets. This attitude created resentment in Nixon himself and suspicion among the administration hard-liners regarding Safire's loyalty. In March 1971, Safire was in such disfavor with Nixon over his outspoken advocacy for the administration's maintaining wide access to the media that he didn't receive a speech assignment or so much as a memo or phone call from the president for three months. In fact, Safire had been wiretapped by the administration soon after the 1968 election, but this didn't stop Nixon from dealing with him. (Safire himself did not discover the wiretapping until mid-1973.)

As happened at Camp David this weekend, the more critical the speech, the more time Nixon spent on it and the smaller the circle of people allowed to see it in advance of its delivery. Usually, the president provided an outline to the speechwriting team that included a crystal-clear structure, with Roman numerals and a logical flow of ideas. Sometimes he would react to a draft initiated by his staff with written comments that went in a much different direction from the material that had been originally suggested. "Once he got hold of the text," recalled former speechwriter Lee Huebner, "it was anyone's guess what would come back." At times, Nixon would write the speech himself, asking for help only with some factual details.

Occasionally, the president acted as a coach to his speechwriters. He once called all of them in to explain that he wasn't being properly understood by the media. He told them that before they submitted anything to him, they should underline in red the three sentences that embodied the ideas on which they wanted the media to focus, what they hoped would be the lead paragraphs and key sound bites. This device proved to be a strong discipline for the writers. Nixon would also emphasize the importance of repetition and the need for extreme brevity so that the networks would find it easy to highlight what he had said. Sometimes, he would provide direct feedback to the author of a particular speech, phoning him after he, Nixon, had delivered the speech to review what had gone well and what hadn't. He was so preoccupied with getting his message across, and so convinced that no one did it well enough, that he often seemed more concerned with the presentation of policy than its substance, a lament expressed by Henry Kissinger and even Haldeman.

For Nixon, a bold speech on a big issue delivered on prime-time TV, one that surprised the public and media (and even many in his own administration), was a major source of personal gratification. Biographer Richard Reeves wrote, "His most important achievements—the remaking of the national and world economy in 1971 and the opening of China in 1972—became public in dramatic television announcements." During his time in office, Nixon gave thirty-seven speeches from the Oval Office.

Understanding Nixon's deepest feelings about televised speeches requires going back to Tuesday, September 23, 1952, when he, aspiring to remain on Eisenhower's ticket as vice president, was embroiled in a scandal related to alleged misuse of political funds. In a desperate effort to avoid a personal and political catastrophe, he went on TV to defend himself with what became known as "the Checkers speech." It was the first nationally televised political speech in the United States and was watched or listened to by some sixty million people, more than a third of the U.S. population at the time. Then vice-presidential candidate Nixon's description of having a modest family effectively tapped into Middle American values. He told the viewers and listeners that the only gift the Nixon family had accepted was a small black-and-white cocker spaniel that his young daughter had named "Checkers." The speech turned out to be a spectacular success. Some surveys showed four million responses in the form of phone calls and letters, virtually all pro-Nixon. A 1999 poll of communications scholars rated the speech among the top six of the twentieth century, including the soaring oratory of FDR, Martin Luther King Jr., and JFK. The speech created a lasting impression in Nixon's mind that he could use TV to end-run the media. He felt he could dig himself out of serious problems by reaching out and communicating directly to the silent majority, and that only television would allow him to do that.

Nixon's penchant for the big TV announcement also embodied his deep suspicion of the media. This resentment went back to his Senate days in the late 1940s and early '50s, when he was attacked by the "liberal eastern press" for identifying State Department official Alger Hiss as a Communist. He hated the way the press portrayed him as a hatchet man for Eisenhower, and he deeply resented the way the media loved JFK. He never forgot the way reporters hung around like hyenas after he lost the race for governor of California in 1962. To Nixon, the press did not just embody a strong liberal bias; it was not just an unelected and unrepresented center of power; it was a mortal enemy. He wanted the press "to be hated and beaten." He would read the daily news summary and personally direct his staff how to respond to unfavorable coverage. When Stuart Loory of the *Los Angeles Times* wrote about the high cost of Nixon's vacation homes, for example, the president told his staff to bar the reporter from entering the White House—an unusually confrontational gesture for that time.

ON SATURDAY MORNING, Connally and the team were having breakfast at Laurel Lodge when Haldeman told them that Nixon had decided to close the gold window and that the speech would be delivered Sunday night.

Many of the Camp David team started the day glancing at newspapers that had been delivered in the early hours. No doubt, they wanted to see if news of their meeting had leaked. That morning, the *Washington Post* had a headline on its front page: "Nixon Economic Advisors Called to Weekend Sessions." The article described a suddenly announced set of meetings at Camp David for Nixon's top economic team, which "led to speculation that the administration's [domestic and foreign economic] problems were undergoing review." To their relief, no details were reported on the substance of the decisions being considered. The

New York Times Business section ran an article, called "Chaotic Trading Weakens Dollar," that talked about the weakening of the dollar and "chaotic trading on world currency exchanges." It mentioned that "on both sides of the Atlantic, there were rumors in the financial community that President Nixon's meeting with his senior economic advisors, scheduled at Camp David in the Maryland mountains this weekend, would produce a United States initiative to shore up the dollar," but gave few other details.

Close to noon, Safire had redrafted the speech and delivered it to Rose Mary Woods to type for Nixon. He then walked over to Laurel for lunch.

CAMP DAVID PROVED an ideal setting for what Nixon wanted to achieve. It was remote, quiet, and personally uplifting. It reminded its visitors of childhood camp, of getting away, of being in simple but comfortable surroundings. It even smelled like camp, especially in summer, when it emitted the woody and earthen fragrances of the surrounding deciduous forest.

A great advantage of this mountain retreat was its relatively small, cozy, and isolated setting. No room for big, unwieldy staffs. No assistants telling you that you were required to leave for another meeting. Shorn of their bureaucratic surroundings, those at the camp could behave more as individuals offering their views on the basis of who they were, what their experiences had been, and what they sincerely believed, rather than having to defend their departments' entrenched positions. Herbert Stein would later recall the elevation, the quiet, the comfortable furnishing, the wide choice of food and recreation, and that everyone was treated as if he were an admiral. He wrote, "The Camp David establishment was arranged to give the participants the sense of their unique value."

Here, too, was a once-in-a-lifetime perk that bolstered the spirits of those who were there. When else and where else could you engage with the president of the United States for hours on end in so intimate and informal a setting? By today's standards, the participants in this meeting were not wealthy men, and the "perk" was even more meaningful than it might be considered today, when presidential advisors are so often people of extraordinary financial means and used to luxury. Connally and Peterson may have been reasonably well off—although, again, not by today's standards—but Arthur Burns, George Shultz, and Paul McCracken had come from academia, and Paul Volcker had been a civil servant or had occupied a research position in a bank for all his professional life.

In the summer of 1971, just twenty-six years after the end of World War II, Camp David was a place where powerful men met to decide the fate of the United States and, by implication, much else in the world. All the participants that weekend understood the dramatic turning point that their decisions represented.

Nixon regularly favored Camp David as his refuge to reach critical decisions or to write his most important speeches. Thus, it was at Camp David that he wrote his eulogy for President Eisenhower in the spring of 1969. It was at Camp David that he decided to send troops into Cambodia in the spring of 1970. Ever the loner, Nixon needed personal space to think through what he would say and do in critical situations. "I never prepared for an important speech or press conference or made a major decision in the oval office," he once wrote. Said Nixon biographer Richard Reeves, "In the mountains, Nixon was forever plotting, planning revolutions great and small, sometimes to build a better world, more often just coups against his own staff and cabinet."

The camp was notable for its low-key and traditional features. The president's lodge had overstuffed furniture and easy chairs with ottomans, often with plaid upholstery. There was a big picture window, a large stone fireplace over which was mounted the Presidential Seal, and a comfortable living room lined with bookshelves and lit by floor lamps rather than overhead lighting. All the cabins, in fact, had large fireplaces made of old stone, wood-beamed ceilings, paneled walls, and old-fashioned kitchenettes. Even in Aspen, everyday china and silverware were the norm.

When Nixon came by helicopter, as he usually did, a decoy chopper often accompanied Marine One to throw off any saboteurs. Security was exceedingly tight and often heavily camouflaged. The airspace was restricted. In 1971, you could not see a front gate, nor any security guards or equipment. Wrote Rear Adm. Michael Giorgione, a former camp commander, "If by accident or intention, you turn in and proceed a few yards down the path, everything changes . . . The silent landscape springs to life in the form of some of the most observant military forces known to man. Here the trees really do have eyes."

In 1971, the security team at the camp was comprised of approximately one hundred specially selected and trained Vietnam marine veterans, who were quartered in barracks on the premises. They guarded the camp's perimeter and the entrance and also manned several watchtowers. When the president or his guests walked or rode golf carts—the usual mode of motor transportation—the marines hid in the woods to keep an eye on them. They would remain stationary, using walkie-talkies to signal back and forth when the dignitaries were passing from one sentry to another. Everyone was given a code name. Nixon was called "The Man," Kissinger was "007." Outside the entry of the camp, the marines had also buried a huge fire hose, attached to a major water source, which they could quickly pull up and use if any large public demonstrations emerged.

AFTER BREAKFAST, THREE working groups met in three separate rooms in Laurel Lodge, the doors of each open to the others. Participation in each group was fluid, and the men wandered into different rooms comfortably and without any fanfare. Without bureaucracies to clear positions, defend turf, and demand legal review at every juncture, the group could be very efficient in comparing ideas and offering refinements. "We were winging it," recalled Safire. "But that was the only way it could have been done, and the quality of the work turned out depended not on the size and depth of the departments but on the intellectual grasps and organizational ability of the men in the cabin."

The organization of the groups was rejiggered that morning. Connally wasn't assigned to any one group but moved among them. The budget and tax group consisted of Shultz; his three deputies, Cap Weinberger, Arnold Weber, and Kenneth Dam; and occasionally Stein and Ehrlichman. Among the issues they focused on were specific tax breaks and spending cuts or postponements. In each case, they did precise calculations of the overall fiscal impact not just for 1972 but the year after. They devoted considerable time to the removal of excise taxes on American automobiles and discussed whether these price reductions would be passed on to consumers. They also reviewed tax breaks for U.S. companies exporting abroad and thereby creating jobs at home, as opposed to those producing goods from subsidiaries abroad and therefore substituting foreign workers for American labor.

The international monetary and trade group was comprised of Burns, Volcker, Peterson, and eventually Bradfield (the Treasury lawyer), with McCracken and Stein episodically involved. Joining them, Connally underlined his feeling that the United States had no choice but to close the gold window, given that gold reserves were so low and, thus, the United States was at the mercy of foreigners. A good deal of time was taken up with the import tax. They went over the same questions that had been raised the day before: How long should the tax last? What would the foreign reaction be? What legal basis should be used? Interwoven in this part of the discussion was how to incorporate the acute problem of the massive penetration of Japanese textiles all over the United States. Burns and Volcker worked out plans for the announcement of closing the gold window and the immediate steps to be taken afterward.

The wage and price controls group consisted of McCracken, Stein, and Weber. Unanimity existed about a wage and price freeze for the first few months, but not much else. What kind of controls would follow? How long would they stay in place? What sorts of exceptions would there be? Who would administer them? What would the penalty be for disobedient institutions? They compared notes on what labor's reaction would be, with Shultz predicting a hostile reception because company dividends and profits would not be controlled, thus giving management a better deal than the unions.

Each group produced a detailed outline of what the nature of the policy decision would be, and eventually all would be rolled up into an overall explanation of the package of initiatives that would become intertwined in one grand policy pronouncement.

LUNCH TOOK PLACE at Laurel around 12:30. During the meal, each working group summarized what it had done. At one point, Herb Stein produced a spoof of an overall fact sheet, which he shared. It read:

On the fifteenth day of the eighth month the President came down from the mountain and spoke to the people on all networks, saying:

"I bring you a Comprehensive Eight Point Program, as follows:

"First, thou shall raise no price, neither any wage, rent, interest, fee or dividend." (In fact, no controls were levied on interest or dividends.)

"Second, thou shalt pay out no gold, neither metallic nor paper.

"Third, thou shall drive no Japanese car, wear no Italian shoe, nor drink any French wine, neither red nor white.

"Fourth, thou shall pay to whosoever buys any equipment ten percent of the value thereof in the first year, but only five percent thereafter." (This referred to investment tax credits.)

"Fifth, thou shall share no revenue and assist no family, not yet." (This referred to postponement of Nixon's two big initiatives, Federal Revenue Sharing and The Family Assistance Plan.)

"Sixth, whosoever buyeth an American automobile, thou shalt honor him, and charge him no tax.

"Seventh, thou shall enjoy in 1972 what the Democrats promised for 1973." (This refers to the fact that most of Nixon's initiatives this weekend were originally ideas of the Democrats.)

"Eighth, thou shall appoint a Council of Elders to consider what to do for an encore."

As lunch ended, Connally organized everyone for the afternoon discussion with Nixon. He suggested four reports: one on the budget, one on wage and price controls, one on international matters, and one to provide a consolidated overview.

WHILE THE GROUPS were working, Nixon had remained at Aspen. He had breakfast at 9:15 a.m. and spent the morning on the phone talking to his daughter Tricia, who was in New York City; to Charles Colson, his special counsel; and at least three times with Kissinger, who was in Washington preparing to fly to Paris for secret talks with the North Vietnamese.

On one of those Saturday morning calls, Nixon told Kissinger that he was planning to give a speech on Sunday night. He mentioned that he would announce the imposition of a surcharge on imports, but he did so in an elliptical way, so that the full political implications of disrupting ties among the allies with no prior consultation didn't immediately register with Kissinger. In his memoirs, Kissinger recalled, "The fact was that a decision of major foreign policy importance had been taken about which neither the Secretary of State nor the National Security Advisor had been consulted." Kissinger alerted his deputy, Gen. Alexander Haig, to keep an eye on what was going on that weekend. Haig in turn told Hormats, Kissinger's chief economic staff person, to be on alert. Hormats immediately tried to contact both Paul Volcker and Shultz's deputy, Kenneth Dam, at Camp David, but he didn't hear back from either that day.

BETWEEN 2:40 P.M. and 6:15 p.m., Nixon met again with the entire team at Aspen Lodge. The import tax again took up a lot of time. Once more, the president expressed his distaste for using the Trading with the Enemy Act as the authority to raise import taxes, although he reversed his position of the day before and allowed an exception for Japanese textiles, but only if it became absolutely necessary. He wasn't even happy with justifying tariffs using the declaration of a balance-of-payments emergency, as would be permitted under GATT. He decided to punt the decision for now, figuring they could come back to it after Sunday's speech. He did say he wanted an across-the-board tariff with very few exceptions.

Sometime that afternoon, Nixon asked to meet with only the top officials without any of their staff. He again brought up closing the gold window. Even though he had made up his mind to do it, and even though he had told Haldeman about his decision, he had not formally told all his top people what he had decided. He wanted to be sure everyone was notified and on board, given the extended and sometimes emotional discussions of the night before, and he might have wanted to reassure himself that no one would break ranks.

THE MEETING ENDED at about 6:15 p.m. From 6:45 to 7:15 p.m., Safire came over to Aspen Lodge to discuss the speech. Nixon said he was generally pleased with the evolving draft but still not comfortable with the language describing the devaluation of the dollar. He worried that most Americans would nevertheless infer a relative diminishment in the value of the greenback. He was concerned about the political impact of being seen as retreating. In reversing positions on wages and prices, in particular— although he had done the same on the dollar–gold link and on free trade—Nixon could open himself up to charges of having betrayed his principles or of having concealed all along what he actually thought. "As I worked with Bill Safire that weekend," Nixon recalled in his memoirs, "I wondered how the headlines would read: Would it be *Nixon Acts Boldly*? Or would it be *Nixon Changes Mind*?"

ABOUT 7:20 P.M., Nixon asked Haldeman to meet him at the pool, where the chief of staff found the president relaxing and smoking a pipe. Nixon wanted to talk about the communications and public relations strategy surrounding his Sunday evening speech.

The rest of the team was having dinner at Laurel and would work late into the evening to further refine the scope and details of the overall program. During their dinner, when Nixon called Ehrlichman to get a read on the mood of the group, his chief domestic advisor reported very high spirits. This report matched what Safire had recorded in his own personal notes when he observed, "The implications of slamming a lid on the American economy was [*sic*] staggering . . . [Yet] it was also more fun than any of the men in the room had ever had in their lifetimes."

That would not have been the case for the treasury undersecretary, though. "Volcker was undergoing an especially searing experience," Safire admitted. "He was schooled in the international monetary system, almost bred to defend it; the Bretton Woods Agreement was sacrosanct to him; all the men he grew up with and dealt with throughout the world trusted each other in crisis to respect the rules and cling to the few constants like the convertibility of gold. Yet here he was participating in the overthrow of all he held permanent; it was not a happy weekend for him."

Later that evening, Safire showed the entire speech to George Shultz and also to Herb Stein, ignoring Nixon's instructions to share each section only with the person most expert in the subject. Burns dropped by Safire's cabin to make a few technical changes, too, asking the speechwriter to remove a negative reference to "international money changers."

Somewhere around 9:00 p.m., Nixon had asked Haldeman, Ehrlichman, and Weinberger, Shultz's deputy, to come over to

Aspen to talk politics. Haldeman described the scene in his diary. "We walked in and the living room was empty. The [president] was down in his study with the lights off and the fire going on in the fireplace, even though it was a hot night out. He was in one of his sort of mystic moods and, after telling us to sit down and informing Cap that this is where he made all his big cognations, he said what really matters here is the same thing as did with Roosevelt, the second Roosevelt. We need to raise the spirit of the country. That will be the thrust and rhetoric of the speech." Nixon said that unless they raised that spirit, everything would fail, but if they raised it, the economy would take off.

Nixon may have been revealing his doubts about America's destiny, or at least his anxiety that the United States was in a heated, economically competitive race for the first time in his life and that it was not entirely clear the country recognized it. "The Japs, Russians, Chinese and Germans still have a sense of destiny and pride, a desire to give their best," he said. "It does not matter when people get out of a race; they lose their spirit, and it can never be recovered. You must have a goal greater than self, either a nation or a person, or you can't be great. Let America never accept being second best. We must try to be what is within our power to be."

The discussion turned back to communication strategy. Nixon wanted Weinberger to brief California governor Ronald Reagan. He again asked Ehrlichman to call Governor Rockefeller. They talked about Shultz and others trying to deal with labor leaders.

At 10:00 p.m., President Nixon was alone. He turned on the TV to watch the Washington Redskins defeat the Denver Broncos in an exhibition football game.

Sometime around midnight, Burns and Safire were walking outside in the cool country air. Burns was reminiscing about the Eisenhower years. He then turned to the weekend, praising Nixon for being a leader who patiently heard all his advisors' views, asked many good questions, and tried hard to cultivate consensus before ultimately deciding. "He's a President now," Burns told Safire. "He has a noble motive in foreign affairs to reshape the world, or at least his motive is to earn the fame which comes from reshaping the world. Who can say what his motive is? But it is moving him in the right direction."

In his diary, Burns recorded his thoughts of the weekend with a sharper, more cynical view. "All in all, there was very little room for any doubt—taking the president's words as he moved from one subject to the next—that he was governed mainly, if not entirely, by a political motive; that he had reached the decision that the kind of changes we are discussing—on prices and wages, taxes, etc.—were essential for the campaign of 1972. If there was any other motive, it either did not come to the surface or I was too occupied with my own thoughts to recognize it."

IN SOME THIRTY-SIX hours, the big decisions had been made, and everyone was on board. Considerable progress was achieved on assembling the background material to support those decisions. The speech announcing the radical new program—filled with dramatic departures from existing policy and encompassing both the domestic and international economies—was almost complete. The plan for consultations and briefings with political, business, and labor leaders was in sight. Tomorrow would be a big day for the United States and the world.

Sunday, August 15

Sunday began for Nixon with a late breakfast with Haldeman. Afterward, he met with the entire team for about ninety minutes, thanking everyone for their dedication. At one point, he leaned over to Haldeman and said that he should make sure that both Connally and Burns got a lot of public credit for the decisions that were made. He then personally orchestrated a picture-taking session, insisting on where everyone should sit or stand. He seemed both ebullient and tense. Amid the air of celebration, he took time to admonish the Filipino orderly for not having enough chairs at hand.

From the pictures themselves, it would be hard to glean that so many big, potentially world-shaking issues had been discussed and resolved over the preceding two days. These men had just decided to inject the U.S. government into the setting of prices and wages across the complex U.S. economy in a way that was unprecedented in American peacetime history. They had agreed to fundamentally restructure Bretton Woods, an agreement that had been in place since the end of World War II and that had been at the heart of astounding global economic progress for over two decades. Yet, looking at the resulting photos, you couldn't see the tension that would weigh on them as they anticipated the task of implementing these decisions and the prospect of extensive negotiations with business, labor, consumers, and several influential governments abroad. Indeed, most everyone looked relaxed and content, as if they had all gone on a hunting or fishing expedition in the wilderness and were now heading home after a vacation. In some of the photos, all the men are smiling—none more so than John Connally.

EVERYONE BUT NIXON, Safire, and Rose Mary Woods left Camp David in two helicopters after lunch at around 2:00 p.m. They were carrying various documents that had been prepared the day before, and now they would have their staffs go over them, filling in or double-checking facts and numbers. Nixon, Safire, and Woods flew back to Washington two and a half hours later.

Aside from a short lunch alone in Aspen Lodge, Nixon spent most of his time with Safire that afternoon. Sitting by the pool, he continued to fine-tune his speech on the subject that was bothering him most: how to describe to the American public the possible devaluation of the dollar. The first challenge was that the subject was complex, as was the case with almost anything to do with currencies. Beyond that, Nixon feared that for the most powerful nation on earth, the notion of presiding over a decline in the dollar's value had a connotation of failure and humiliation. After all, it meant that the money every American had in his or her hand or bank would be worth less tomorrow in terms of other currencies. It signaled that foreign products—as measured by rising levels of imports—would become more expensive, thus making Americans poorer. How to present the diminishment of the dollar in a positive way?

With Safire by his side, Nixon went over every word, line by line, softening much of the rhetoric. For example, he changed "tough" competition to "strong" competition, "the American people" to "every American." By the time it was delivered, the speech had undergone four complete drafts.

Nixon also came up with a slogan for describing the overall package; he would call it the "New Economic Policy," later abbreviated to NEP. It was a simple expression, because in Nixon's mind the dramatic substance would speak for itself. When Safire heard it, he felt vaguely uncomfortable, as if he thought it had been used before. It took him a while to recall that Vladimir Lenin had also called his program the New Economic Policy. The name nevertheless remained unchanged. The media had too much else to focus on for the title to have much impact anyway.

On the short helicopter flight back to Washington, Nixon rehearsed his speech, with Safire and Woods listening. At one point, he told Safire, "You know when all this was cooked up? Connally and me, we had it set sixty days ago." That was a slight exaggeration. In terms of the president and treasury secretary having a detailed discussion, late July seems more likely, although it was true that Treasury had been making contingency plans for a dollar crisis for over two years. Nixon may also have convinced himself that the radical U-turn he was taking was based on thorough analysis and planning, rather than knee-jerk impulses. Safire sensed that Nixon was making these decisions reluctantly because they went against his free-market inclinations, especially the wage and price aspects. Nevertheless, it was clear he loved putting his opponents off balance. "Nixon hated to do it," recalled Safire, "but he loved doing it."

Meanwhile, back in Washington, the staffs of Treasury, OMB, the Federal Reserve, the NSC, and the White House were scrambling to get everything in order for the 9:00 p.m. speech. Among the tasks at hand was the need to notify key officials abroad in time not to insult them more than necessary, but not far enough in advance to elicit negative reactions before the speech was actually given.

Burns contacted his key central bank counterparts by phone or telex. Because of decades of close cooperation, the Fed knew how to operate clearly and efficiently with its counterparts. Not so with regard to the administration. NSC staffer Hormats, having placed calls on Saturday, finally heard back from Volcker and Dam. He was familiar with most of the issues; he had just never imagined they would come to a head so soon. Also, he was shocked at the decision to implement a surcharge on imports. It was not a subject he had heard being seriously discussed, and to him this was a bridge too far for a nation that had been a champion of free trade.

After getting a quick summary of what had been decided, Hormats asked whether a plan existed to alert heads of state and foreign ministers. Hearing that nothing had yet been done, he volunteered to draft cables. Lacking enough details, he needed help from Volcker or Dam, then passed the cables through the State Department, which would then send them through its secure communications network. The cable from the president to West German chancellor Willy Brandt was illustrative. It started with the measures that Nixon was going to announce shortly. It then asked for cooperation on the monetary front in very general terms and noted that John Connally would be in touch with West Germany's finance minister and that Paul Volcker would travel soon to Western Europe for face-to-face consultations. Similar cables went out to the leaders in London, Paris, Rome, and Tokyo.

Another person who was scrambling that afternoon was John Petty, assistant secretary of the treasury for international affairs, who was holding down "base camp" at the Treasury over the weekend. It was Petty and his team who did the bulk of the work behind Treasury's contingency plans and who would be preparing additional briefing materials surrounding the speech.

On this Sunday afternoon, Petty was checking the consolidated fact sheet that Paul Volcker had given him and that would be delivered to the media shortly before Nixon's speech. He was also overseeing documentation relating to the need to notify the International Monetary Fund that the United States would be closing the gold window. Petty passed the draft for the IMF to Volcker, who forwarded it to Connally. Sometime in the early evening, Connally sent that notice to Pierre-Paul Schweitzer, managing director of the IMF. It was a historic and dramatic communication, saying that this letter superseded the letter of May 20, 1949, signed by Treasury Secretary John Snyder. From now on, Connally wrote, the United States would no longer freely buy and sell gold for settlement of transactions under the IMF Articles of Agreeement. He nevertheless pledged to work with the IMF to promote exchange rate stability and to avoid countries' competing with one another by artificially lowering their rates, otherwise called "competitive depreciations."

At about 5:00 p.m., Nixon called John Mitchell, his attorney general and close confidant. The president was effusive. "It's quite a bundle," he told Mitchell. "It's going to have quite an impact . . . It's really going to pull the rug out from underneath everyone concerned."

Early in the day, Haldeman's team had notified key media officials that the president would be making a major announcement and that they should hold space and broadcast times for that evening and tomorrow morning. He invited the media to an

8:15 p.m. pre-speech press conference in the East Room of the White House. As each reporter arrived, he or she was handed a document called "Explanatory Material of the President's Economic Program." Consisting of twenty-two single-spaced pages, it was a detailed summary of what Nixon's men had been working on these past two days.

RON ZIEGLER, THE White House press secretary, informed reporters that they were to remain in the press room until the speech, scheduled to begin at 9:00 p.m., was over. TV sets would be brought in for them to watch the speech, and texts of the speech would be given out after Nixon was finished. In those pre–internet and cell phone days, those in the room would have had no way to communicate with anyone outside the room.

Connally, Shultz, and McCracken were seated up front, facing the press. Connally began the media briefing with a broad overview. One theme was that the new program was a long time in the making. Another was its bold across-the-board approach. "The President has for a long time been considering what comprehensive, integrated course of action he might take, looking toward the solution to all the problems that beset the country domestically and foreign," the treasury secretary said.

Questions from the media immediately went to the heart of the matter. Isn't the president fundamentally changing course? What happened in the last several weeks that made him go in this new direction? Connally blamed the about-face on the pressure from foreign markets, prompting more questions about what would happen to the value of the dollar. Connally: "I will be perfectly frank with you, none of us know for certain what will occur." What will happen to gold? What happens after the wage-price freeze? At one point, Shultz jumped in to take the heat off any one issue: "I think the important thing to stress here is the set of things that is put forward and their interrelationship. It is not so much any one thing but the whole set of things." Connally and Shultz were forced to dance around a number of issues. At one point, Connally said, "I don't think either [Shultz] or I ever said we were not for a price freeze. He and I said we were not for wage and price controls. As I interpret those two phrases, they mean entirely different things." Both Connally and Shultz were, of course, being disingenuous. They knew that the freeze would have to be followed by some yet-to-be-determined type of controls, or else wages and prices would snap back to where they were and maybe even rise in order to make up for lost time. The press conference ended at 8:50 p.m.

Just before the speech, a number of calls took place from the White House staff alerting key people that Nixon would be giving a major speech and, in some cases, giving a heads-up on the key points. For example, Marina von Neumann Whitman, who was a staff member of the CEA and who wrote many memos for her boss, Paul McCracken, had no idea what had transpired at Camp David. On Sunday morning she received a call from McCracken's assistant, Sidney Jones, asking her to watch Nixon's speech that night. (Under President Ford, Whitman would become the first woman member of the CEA.) Whitman relayed the phone conversation to her husband, repeating Jones's words that "Nixon was going to drop a bomb." Upon hearing this, her young daughter asked whether the bomb would drop right there in the kitchen. That Sunday afternoon, Herb Stein called Alan Greenspan. "The President wants you to listen to the speech," Stein said, and gave Greenspan a few highlights. Greenspan recalls that when he hung up, he reached for something and wrenched his back, causing pain for several weeks. He half-jokingly speculated that what he had heard was such a surprise that it may have caused the injury.

(Nixon would eventually recommend Greenspan for chairman of the CEA, a position he took right after Nixon resigned, and Greenspan would later become chairman of the Federal Reserve in the Reagan administration, serving in that position over almost two decades.)

Minutes before the speech, Secretary Rogers called Prime Minister Eisaku Satō of Japan to warn him about what the president would say. Paul Volcker called Yasuke Kashiwagi, senior advisor to the Ministry of Finance, as well. American officials were acutely aware of Japanese sensitivities to surprises from Washington. After all, the United States had become Japan's closest and most vital ally, its military protector, and its closest friend in the world. Even the U.S. embassy in Tokyo alerted the Japanese ministries and the Bank of Japan. Someone in the Japanese finance ministry relayed the U.S. embassy's alert to Toyoo Gyohten, a special assistant to the vice minister of finance for international affairs. Gyohten immediately proceeded to search for a shortwave radio so that he could hear Nixon's speech on Voice of America (these being pre-CNN days). Looking back on the experience, he remembered, "We called this the Nixon Shock. All of Japan was caught by surprise and it was a big one." (In 1986, Gyohten became a highly visible and respected vice minister of finance for international affairs.)

At 8:35 p.m., Nixon sat with his makeup consultant, Ray Voege. Less than twenty minutes later, the president walked over to the Oval Office with his special assistant, Mark Goode. At 9:00 p.m., the three television networks all broke into regular programming to broadcast the speech. In prior discussions with his staff, Nixon had expressed concern about interrupting NBC's wildly popular program *Bonanza*, the second-longest-running Western on U.S. television.

DRESSED IN A gray suit, white shirt, and gray tie, Nixon was seated at his desk with the American flag on his right, a presidential flag on his left, and a greenish-gray curtain behind him. The president was never known for his formal speeches; audiences saw a more relaxed and effective orator when Nixon was speaking off the cuff. True to form, he gave a stiff performance. He spoke without a teleprompter, holding a sheath of seventeen triple-spaced pages. As he read, he peeled one off after another, his eyes frequently glancing at the text. His copy of the speech was structured in outline form, with lots of white space and big breaks to separate major sections. He stumbled over several words, and at one point he wiped what seemed to be a bead of sweat below his nose with the back of his hand. He appeared especially uncomfortable when talking about the dollar. Occasionally, the lighting bounced off his receding hairline, creating a distracting glare. (Nixon would be critical of the whole setup in his subsequent discussions with Haldeman later that week.) The president's one slight smile came when he said we would make no friends among foreign money traders. The camera zoomed in on his face for most of the time, then zoomed out as he was finishing off.

Below, in italics, is the text, followed by my comments.

Address to the Nation Outlining
a New Economic Policy: The Challenge of Peace

August 15, 1971

Good evening:

I have addressed the Nation a number of times over the past 2 years on the problems of ending a war. Because of the progress we have made toward achieving that goal, this Sunday evening is an appropriate time for us to turn our attention to the challenges of peace.

America today has the best opportunity in this century to achieve two of its greatest ideals: to bring about a full generation of peace, and to create a new prosperity without war.

Nixon began with a big theme that linked one of his main goals, ending the war in Vietnam, with the need to address his greatest liability, the deteriorating economy.

This not only requires bold leadership ready to take bold action—it calls forth the greatness in a great people.

Prosperity without war requires action on three fronts: We must create more and better jobs; we must stop the rise in the cost of living; we must protect the dollar from the attacks of international money speculators.

Here was a clear road map for how Nixon would frame the issues to the American people. He led with jobs, the key political challenge. He understood the dangers of inflation, but he never saw it as a prominent election issue. And he described the challenge of the dollar as though America were being attacked—an us-versus-them battle.

We are going to take that action—not timidly, not half-heartedly, and not in piecemeal fashion. We are going to move forward to the new prosperity without war as befits a great people—all together, and along a broad front.

The sheer scope of the actions being proposed and the relationship among the various initiatives was a theme the administration wanted to emphasize—and not only because Nixon relished big, bold proposals. By saying all the actions fit together, the president and Connally believed that it would be harder for critics to attack any one part. Also, if everything seemed interconnected, the supporters of any one component would feel they had to get behind the entire program in order to save the portion they cared most about. Indeed, Nixon was offering something for everyone. Liberals would like the wage and price constraints. Conservatives would like the investment incentives for business. Most everyone would support closing the gold window, which they would equate with a cheaper dollar, which would mean more exports and fewer imports, which in turn could mean higher employment. "Without visible regret, certainly without apology, [Nixon] moved across a broad front in a series of actions that changed the economic landscape of the world," Safire wrote in his account of the weekend.

> *The time has come for a new economic policy for the United States. Its targets are unemployment, inflation, and international speculation. And this is how we are going to attack those targets.*
>
> *First, on the subject of jobs. We all know why we have an unemployment problem. Two million workers have been released from the Armed Forces and defense plants because of our success in winding down the war in Vietnam. Putting those people back to work is one of the challenges of peace, and we have begun to make progress. Our unemployment rate today is below the average of the 4 peacetime years of the 1960's.*
>
> *But we can and we must do better than that.*

In attributing unemployment to the end of the Vietnam War, Nixon was laying the blame for America's problems on a war he hadn't started, thus deflecting responsibility for economic failures on his watch.

The time has come for American industry, which has produced more jobs at higher real wages than any other industrial system in history, to embark on a bold program of new investment in production for peace.

To give that system a powerful new stimulus, I shall ask the Congress, when it reconvenes after its summer recess, to consider as its first priority the enactment of the Job Development Act of 1971.

By leading with jobs and other incentives for investment, Nixon hoped to get the broadest array of domestic constituencies on board at the outset. A good deal of the Camp David discussion over the past two days had concerned precisely what measures the administration would propose to reduce unemployment. In an earlier draft, "Investment Tax Incentives" was the title of the new legislation that would be proposed for these tax breaks. Nixon pressed for something that would better resonate with the average American. Safire came up with "Job Development."

I will propose to provide the strongest short term incentive in our history to invest in new machinery and equipment that will create new jobs for Americans: a 10 percent Job Development Credit for 1 year, effective as of today, with a 5 percent credit after August 15, 1972. This tax credit for investment in new equipment will not only generate new jobs; it will raise productivity; it will make our goods more competitive in the years ahead.

Second, I will propose to repeal the 7 percent excise tax on automobiles, effective today. This will mean a reduction in price of about $200 per car. I shall insist that the American auto industry pass this tax reduction on to the nearly 8 million customers who are buying automobiles this year. Lower prices will mean

that more people will be able to afford new cars, and every
additional 100,000 cars sold means 25,000 new jobs.

The administration applied this tax reduction only to American cars, not imported ones. Foreign governments considered this "Buy American" policy discriminatory and a violation of trade agreements, and would vehemently oppose it in subsequent negotiations.

Third, I propose to speed up the personal income tax exemptions scheduled for January 1, 1973, to January 1, 1972—so that taxpayers can deduct an extra $50 for each exemption 1 year earlier than planned. This increase in consumer spending power will provide a strong boost to the economy in general and to employment in particular.

The tax reductions I am recommending, together with this broad upturn of the economy which has taken place in the first half of this year, will move us strongly forward toward a goal this Nation has not reached since 1956, 15 years ago: prosperity with full employment in peacetime.

The return to postwar prosperity was a theme Nixon would sound again and again through the rest of his administration. However, such prosperity without inflation proved a bridge too far over the 1970s. This disappointment was due, in large part, to the unique circumstances that existed between the late 1940s and the late 1960s, circumstances that could never be duplicated. After all, those years had been characterized by the unlimited appetite in Western Europe and Japan for American goods during their recovery from the war; the lack of foreign competition for U.S. firms during most of that time; increasing productivity in the U.S. workforce in the early postwar years; and the powerful rise of a military-industrial complex that fueled production, jobs, and new technologies across the nation. Few of these factors continued into the 1970s.

> *Looking to the future, I have directed the Secretary of the Treasury to recommend to the Congress in January new tax proposals for stimulating research and development of new industries and new techniques to help provide the 20 million new jobs that America needs for the young people who will be coming into the job market in the next decade.*

Nixon made these proposals, but it did not appear his heart was in them. Peter Peterson had been a big advocate of increased government support for industries of the future. He pushed for a greater emphasis on structural issues, such as extensive programs for retraining workers for higher-skilled occupations, but they were never taken up within the administration.

> *To offset the loss of revenue from these tax cuts which directly stimulate new jobs, I have ordered today a $4.7 billion cut in federal spending.*

Nixon was acutely concerned about not being seen as a big spender and as adhering to the Republican doctrine of keeping the budget deficit under control. He calculated that the Democratic-controlled Congress would not implement these cuts, and then the deficits would be their fault, not his. This would allow him to wage his next presidential campaign against the fiscally irresponsible Democrats.

> *Tax cuts to stimulate employment must be matched by spending cuts to restrain inflation. To check the rise in the cost of*

*Government, I have ordered a postponement of pay raises and a
5 percent cut in Government personnel.*

I have ordered a 10 percent cut in foreign economic aid.

This cut in foreign aid had not been a big subject at Camp
David. With Kissinger and Rogers absent from the meeting, almost
no discussion of foreign policy took place. Rogers had approved
the foreign aid cut sometime over the weekend, perhaps by phone.

*In addition, since the Congress has already delayed action
on two of the great initiatives of this Administration, I will ask
Congress to amend my proposals to postpone the implementation
of revenue sharing for 3 months and welfare reform for 1 year.*

Nixon again made a virtue of necessity, for he knew that these
programs would have been delayed by Congress anyway.

*In this way, I am reordering our budget priorities so as to con-
centrate more on achieving our goal of full employment.*

*The second indispensable element of the new prosperity is to
stop the rise in the cost of living.*

While inflation had been coming down, it was still too high.
The combination of unemployment and inflation was the single-
biggest domestic challenge to the administration. Economists
simply had no remedy for "stagflation."

*One of the cruelest legacies of the artificial prosperity pro-
duced by war is inflation. Inflation robs every American, every
one of you. The 20 million who are retired and living on fixed
incomes—they are particularly hard hit. Homemakers find it
harder than ever to balance the family budget. And 80 million*

*American wage earners have been on a treadmill. For example,
in the 4 war years between 1965 and 1969, your wage increases
were completely eaten up by price increases. Your paychecks were
higher, but you were no better off.*

*We have made progress against the rise in the cost of living.
From the high point of 6 percent a year in 1969, the rise in consumer prices has been cut to 4 percent in the first half of 1971.
But just as is the case in our fight against unemployment, we can
and we must do better than that.*

*The time has come for decisive action—action that will break
the vicious circle of spiraling prices and costs.*

*I am today ordering a freeze on all prices and wages throughout
the United States for a period of 90 days. In addition, I call upon
corporations to extend the wage-price freeze to all dividends.*

The implications of injecting Washington so extensively into
the everyday machinery of American capitalism—in peacetime
no less—was breathtaking, especially for a Republican. Wage
and price restraints became the key to everything else. After all,
the Job Development part of the program could be inflationary,
as would be increased tariffs and a depreciating dollar. Without
wage and price controls, there would have been no way to control
the inflationary pressures of the other parts of the package.
New York Times journalist Leonard Silk wrote that the public
had been yearning for action on wages and prices and that
"[i]mposing a freeze on wages and prices dramatized the president's
anti-inflation policy and forced it deeply into the consciousness
of consumers, workers and businessmen."

An analogy could be made to Nixon's China policy. It has
been said that only Nixon could have extended a hand to China
in 1971 because of his indisputable anticommunist credentials

going back decades. When it came to wage and price controls, he had opposed them for so long that his sudden embrace of them undoubtedly brought far less criticism than would have been the case if a Democrat had embraced the same policy.

> *I have today appointed a Cost of Living Council within the Government. I have directed this Council to work with leaders of labor and business to set up the proper mechanism for achieving continued price and wage stability after the 90-day freeze is over.*
>
> *Let me emphasize two characteristics of this action: First, it is temporary. To put the strong, vigorous American economy into a permanent straitjacket would lock in unfairness; it would stifle the expansion of our free enterprise system. And second, while the wage-price freeze will be backed by Government sanctions, if necessary, it will not be accompanied by the establishment of a huge price control bureaucracy. I am relying on the voluntary co-operation of all Americans—each one of you: workers, employers, consumers—to make this freeze work.*

The policies and actions that would come after the freeze would be among the biggest questions raised by the speech. In this passage, Nixon was being misleading at best, using words such as *temporary*, saying there would be no "bureaucracy," and generally giving the impression that the entire enterprise would be short and handled with a very light government touch. On the one hand, he was expressing great reservations about the entire project, emphasizing the temporary nature of the freeze and his aversion to building a bureaucracy. On the other hand, he was establishing a Cost of Living Council, which, from the broad nature of its name, augured a large-scale government intervention in the American economy for a long time to come.

Working together, we will break the back of inflation, and we will do it without the mandatory wage and price controls that crush economic and personal freedom.

This turned out to be untrue. Virtually everything Nixon detested about wage and price controls came to pass.

The third indispensable element in building the new prosperity is closely related to creating new jobs and halting inflation. We must protect the position of the American dollar as a pillar of monetary stability around the world.

In the past 7 years, there has been an average of one international monetary crisis every year. Now who gains from these crises? Not the workingman; not the investor; not the real producers of wealth. The gainers are the international money speculators. Because they thrive on crises, they help to create them.

Nixon struggled to explain currency issues to the public. In the lead-up to Camp David, he considered more than once having Connally himself deal with this aspect of the program. But in the end, he had no choice but to address it this evening. First, the dollar had been under sustained attack from abroad, and currency issues were in the news. Most urgently, Washington was afraid of a foreign run on gold. Under no circumstances would it allow its gold stock to be depleted, and so, closing the gold window seemed imminent. Better to preempt that move and make it seem as if the United States were in charge, so the Nixon-Connally logic went.

Also significant was Nixon's use of the term "international money speculators." It was a choice of words that Burns, in particular, objected to and unsuccessfully tried to have removed from the speech. In labeling foreign dollar holders this way, Nixon created a clear-cut enemy. The United States would not stand for such aggression and would pull out all the stops to defend its interests, the president implied. "Nixon was at his best when he was on the attack," observed former Nixon speech-writer Lee Huebner. "He liked having an enemy." The theme of these underhand foreigners continued in the next paragraphs.

> *In recent weeks, the speculators have been waging an all-out war on the American dollar. The strength of a nation's currency is based on the strength of that nation's economy—and the American economy is by far the strongest in the world. Accordingly, I have directed the Secretary of the Treasury to take the action necessary to defend the dollar against the speculators.*
>
> *I have directed Secretary Connally to suspend temporarily the convertibility of the dollar into gold or other reserve assets, except in amounts and conditions determined to be in the interest of monetary stability and in the best interests of the United States.*

This move was the beginning of a radical reform of Bretton Woods, although it wasn't presented that dramatically.

> *Now, what is this action—which is very technical—what does it mean for you?*
> *Let me lay to rest the bugaboo of what is called devaluation.*

It is unclear why Nixon talked so nakedly about devaluation, because in his press conference of an hour before, Connally had said that neither he nor anyone else knew precisely what would happen when the gold window was closed. Nixon could have focused on the need for other nations to strengthen their currencies, as their keeping them artificially weak was to the detriment of American citizens. At Camp David, a discussion of the political implications of a devaluation was on everyone's mind. When the United Kingdom or France had devalued, the government in each case was accused of having surrendered to the markets. Devaluation had become a political humiliation, and Nixon, determined to avoid this negative stigma, moved to get ahead of it and frame it in his own combative way.

> *If you want to buy a foreign car or take a trip abroad, market conditions may cause your dollar to buy slightly less. But if you are among the overwhelming majority of Americans who buy American-made products in America, your dollar will be worth just as much tomorrow as it is today.*

Here the president was playing loose with the truth. A cheaper dollar would make imports more expensive, and imports were beginning to weave themselves into all aspects of American production. Also, even if imports were impeded, the result would be less competition in America, allowing domestic producers to raise their prices more easily. So, a dollar that was cheaper than it had been vis-à-vis other countries was not without its significant costs.

> *The effect of this action, in other words, will be to stabilize the dollar.*

The word *stability* was also misleading. The administration wanted a weaker dollar and a stronger mark and yen—big changes in the existing monetary order that would lead to uncertain currency movements all over the world.

Now, this action will not win us any friends among the international money traders. But our primary concern is with the American workers, and with fair competition around the world.

Again, Nixon was creating an enemy.

To our friends abroad, including the many responsible members of the international banking community who are dedicated to stability and the flow of trade, I give this assurance: The United States has always been, and will continue to be, a forward-looking and trustworthy trading partner. In full cooperation with the International Monetary Fund and those who trade with us, we will press for the necessary reforms to set up an urgently needed new international monetary system. Stability and equal treatment is [sic] in everybody's best interest. I am determined that the American dollar must never again be a hostage in the hands of international speculators.

Here was Nixon's attempt to say that America still supported a cooperative approach to the international economy. The United States was not proposing to shatter or abandon the arrangements established and led by Washington at Bretton Woods, but to change some of the relationships within them. Nixon was signaling that the United States still believed in free trade and free flows of capital and that it supported the IMF rules-based monetary system. It's just that the United States has been treated unfairly within the Bretton Woods framework and that there thus must be some adjustments. Many questions remained unanswered. Washington was reducing the burdens of its global leadership, but how far would it go? The United States had not consulted with its allies. So, to what degree would it now push unilaterally and how much would it be open to genuine multilateral negotiations?

> *I am taking one further step to protect the dollar, to improve our balance of payments, and to increase jobs for Americans. As a temporary measure, I am today imposing an additional tax of 10 percent on goods imported into the United States. This is a better solution for international trade than direct controls on the amount of imports.*

The imposition of an import tax had proven particularly controversial at Camp David, but ultimately, Nixon and Connally wanted it for negotiating leverage. The tariff would become even more contentious among the allies than closing the gold window. Western Europe and Japan were thriving partly because the United States was consuming so much of their imports. A tariff that raised the price of their export sales to the United States was seen as a major threat to their economies and their future. In addition, it had been the United States that led successive global trade negotiations to bring tariff levels down. Free trade was seen as a cornerstone of U.S. ideology and policy. By using trade as a weapon, Washington was now moving in the opposite direction.

> *This import tax is a temporary action. It isn't directed against any other country. It is an action to make certain that American products will not be at a disadvantage because of unfair exchange rates. When the unfair treatment is ended, the import tax will end as well.*

Nixon here used the word *temporary* again, just as he had with wage and price controls, implying that the tariff increase would be removed when unfair treatment of the United States ended. However, the criteria for removal were vague, and the import surcharge definitely had a gun-to-the-head feeling—which was exactly what Nixon and Connally wanted.

> *As a result of these actions, the product of American labor will be more competitive, and the unfair edge that some of our foreign competition has will be removed. This is a major reason why our trade balance has eroded over the past 15 years.*

Nixon used the word *unfair* for a second time. Americans had been exceedingly generous to Western Europe and Japan, and Nixon implied they were now being betrayed. Not surprisingly, the notion that American policies themselves were partly responsible for these outcomes—especially lax budgets and easy monetary policy—was absent from the speech. The following paragraphs further underlined the theme that foreigners were to blame.

> *At the end of World War II the economies of the major industrial nations of Europe and Asia were shattered. To help them get on their feet and to protect their freedom, the United States has provided over the past 25 years $143 billion in foreign aid. That was the right thing for us to do.*
>
> *Today, largely with our help, they have regained their vitality. They have become our strong competitors, and we welcome their success. But now that other nations are economically strong, the time has come for them to bear their fair share of the burden of defending freedom around the world. The time has come for exchange rates to be set straight and for the major nations to compete as equals. There is no longer any need for the United States to compete with one hand tied behind her back.*

This paragraph crystalized ideas that had been building for a decade and were now planted deep in the administration's psyche and worldview: The world has changed. The United States cannot and will not shoulder all the burdens it once did, and others must step up and pitch in more. There will have to be a big international adjustment of responsibilities in the world economy, akin to the U.S. foreign policy announced in the 1969 Nixon Doctrine and in line with the Peterson Report, "The United States in the Changing World Economy."

> *The range of actions I have taken and proposed tonight—on the job front, on the inflation front, on the monetary front—is the most comprehensive new economic policy to be undertaken in this Nation in four decades.*
>
> *We are fortunate to live in a nation with an economic system capable of producing for its people the highest standard of living in the world; a system flexible enough to change its ways dramatically when circumstances call for change; and, most important, a system resourceful enough to produce prosperity with freedom and opportunity unmatched in the history of nations.*
>
> *The purposes of the Government actions I have announced tonight are to lay the basis for renewed confidence, to make it possible for us to compete fairly with the rest of the world, to open the door to new prosperity.*
>
> *But government, with all of its powers, does not hold the key to the success of a people. That key, my fellow Americans, is in your hands.*

As Nixon drew to his conclusion, he returned to the core Republican philosophy of individual responsibility, individual initiative, and the need to reduce government involvement in everyday life. The irony was that so much of this speech went in the opposite direction, making the case for much government intervention. In fact, Nixon was catering to a great American myth that glorifies individualism and ignores the critical positive role that government plays in our society.

> *A nation, like a person, has to have a certain inner drive in order to succeed. In economic affairs, that inner drive is called the competitive spirit.*
>
> *Every action I have taken tonight is designed to nurture and stimulate that competitive spirit, to help us snap out of the self-doubt, the self-disparagement that saps our energy and erodes our confidence in ourselves.*
>
> *Whether this Nation stays number one in the world's economy or resigns itself to second, third, or fourth place; whether we as a people have faith in ourselves, or lose that faith; whether we hold fast to the strength that makes peace and freedom possible in this world, or lose our grip—all that depends on you, on your competitive spirit, your sense of personal destiny, your pride in your country and in yourself.*
>
> *We can be certain of this: As the threat of war recedes, the challenge of peaceful competition in the world will greatly increase.*
>
> *We welcome competition, because America is at her greatest when she is called on to compete.*
>
> *As there always have been in our history, there will be voices urging us to shrink from that challenge of competition, to build a protective wall around ourselves, to crawl into a shell as the rest of the world moves ahead.*

Again, the Nixon administration didn't see itself as protectionist. It was not saying that globalization is bad. To the contrary, it was struggling with how to maintain the balance between globalization and stability at home.

> *Two hundred years ago a man wrote in his diary these words: "Many thinking people believe America has seen its best days." That was written in 1775, just before the American Revolution— the dawn of the most exciting era in the history of man. And today we hear the echoes of those voices, preaching a gospel of gloom and defeat, saying the same thing: "We have seen our best days."*
>
> *I say, let Americans reply: "Our best days lie ahead."*
>
> *As we move into a generation of peace, as we blaze the trail toward the new prosperity, I say to every American: Let us raise our spirits. Let us raise our sights. Let all of us contribute all we can to this great and good country that has contributed so much to the progress of mankind.*
>
> *Let us invest in our Nation's future, and let us revitalize that faith in ourselves that built a great nation in the past and that will shape the world of the future.*
>
> *Thank you and good evening.*

ADMINISTRATION ESTIMATES SHOWED that 46,200,000 Americans tuned in to the three networks (ABC, CBS, and NBC) to watch Nixon. That was about a quarter of the U.S. population.

Connally watched the speech in his spacious office at the Treasury, along with Volcker and his assistant secretary, John Petty. Joining them at Connally's invitation was Pierre-Paul Schweitzer, managing director of the IMF. Schweitzer, a hero of the French Resistance during the war, not to mention a survivor of the Buchenwald concentration camp, was already agitated when he was invited at the last minute to witness an announcement he knew nothing about, one that turned out to be so central to his professional responsibilities. The four men watched the speech in silence. As soon as Nixon finished, Connally received a call and hurriedly excused himself. Volcker had started to brief Schweitzer when he, too, was called out. Petty, embarrassingly contrite, asked Schweitzer if he wanted to discuss the speech, but a humiliated Schweitzer made a polite but abrupt exit. The incident portended rough waters for the United States–IMF relationship in the months ahead.

Years later, Volcker reflected on his own reaction to the speech. At the beginning of the weekend, he had given Connally a proposed draft. "I suppose [what I wrote] was a typical devaluation speech, aimed at calming financial markets and reassuring central bankers," he wrote. "It included a few of what I thought were necessary mea culpas, along with promises to maintain internal discipline, deal with inflation, and work cooperatively to reform and improve the monetary system." Volcker went on to explain what a different tack Nixon took. "Lo and behold, the suspension of gold payments, something that I had spent so much of my working life defending, became a bold new initiative. The international crisis provided a plausible rationale for what otherwise might have been criticized as an abrupt and embarrassing change in domestic policy," particularly, he said, with regard to wage and price restraints and tax reductions. "I learned a good lesson about what masterful politicians do," Volcker said.

IMMEDIATELY FOLLOWING THE speech, the three major networks provided short commentary for about ten to fifteen minutes each. Most notably, ABC's Bill Gill, joined by congressional correspondent Bob Clark, interviewed the chairman of the Council of Economic Advisers, Paul Mc-Cracken. They posed rapid-fire questions about what would come next for the dollar, for tariffs, for wages and prices, catching McCracken off guard and forcing him to resort to responses such as "We'll have to see." Their back-and-forth also involved speculation over whether the existing policies had failed; to what extent foreign policy consequences had been taken into account; whether the purpose of the new initiatives was to bolster the president's 1972 reelection prospects; and whether the president had stolen the Democrats' ideas and left them no room to do anything but applaud. Gill pressed McCracken on what all this would mean for the average American. That same theme pervaded the CBS evaluation, led by Robert Pierpont and his colleague Daniel Schorr. This was perhaps the most intelligent discussion, one that included a simple but accurate description of how the international situation was responsible for so many of the pressures on the United States and an explanation of why Nixon had felt compelled to act. Given the short time to prepare, the three networks did a remarkably thoughtful job of succinctly summarizing and explaining what Nixon had said, as well as raising the key unanswered questions.

Meanwhile, Nixon received several congratulatory calls from his senior advisors. In addition to George Shultz, he heard from Clifford Hardin, secretary of agriculture, who would have had his eyes on more open markets abroad for U.S. farm products; Labor Secretary James Hodgson, who would have to play a big role in wage and price issues, even though he had been overshadowed by Shultz and cut out of much that had already been decided; and George Romney, head of Housing and Urban Development (HUD), who had been a big proponent within the administration of wage and price controls. Burns called, too, as did Kissinger's deputy, Al Haig.

RIGHT AFTER THE speech, Volcker prepared to fly to London in order to be ready the next day to brief finance ministers on the biggest change in international finance since Bretton Woods. He knew he would be grilled on many of the uncertain issues—how long the tax surcharge would last, what precisely the United States was looking for with regard to exchange rates, when the gold window would be reopened, and the details of many of the domestic programs. He knew he wouldn't be able to answer most of these questions because, in fact, the policies had not been determined. So, he would have to temporize and say he was there to collect foreign views to bring back to Washington for input into future decisions.

Sometime Sunday evening, Paul Volcker received a note from Haldeman's office saying the military aircraft would be ready at Andrews Air Force Base in Maryland at 11:00 p.m. It would have four bunks on it and food for the evening and next morning. It would be available to shuttle Volcker and his two-man team around Western Europe as required and would return at midnight Wednesday, August 18. Unfortunately for Volcker, his six-foot, seven-inch frame was too big for the bed, and he had to sleep in the aisle. Volcker's trip would be the beginning of four months of intense international negotiations.

IV.

THE FINALE

A meal at Laurel Lodge at Camp David. Secretary of the Treasury John Connally, his back to us, is flanked by Deputy Director of the Office of Management and Budget Caspar Weinberger (*left*) and Federal Reserve Board Chairman Arthur Burns (*right*).

Nixon's team at the end of the weekend. *Left to right*: Treasury Undersecretary Paul Volcker, Assistant to the President Peter Peterson (*behind Volcker*), Arthur Burns, Council of Economic Advisers member Herb Stein, Council of Economic Advisers Chairman Paul McCracken, John Connally, and Office of Management and Budget Director George Shultz. The second row, aside from Peterson, is senior staff.

Paul Volcker (*left*), with John Connally

Peter Peterson (*left*), with George Shultz

Left to right: Peter Peterson, White House speechwriter William Safire,
John Connally, Herb Stein, and Paul McCracken

Nixon's chief of staff, H. R. Haldeman (*foreground*), with George Shultz

William Safire (*facing*), with Herb Stein

Arthur Burns

John Connally

President Richard Nixon and John Connally

The television studio minutes before President Nixon's August 15, 1971, address

The Aftermath

In the days after the speech, the reaction in the United States was, with a few exceptions, wildly supportive. On August 16 the stock market jumped 32 points, a 3 percent gain and the largest one-day jump in its history up to then. Over the following week, White House aides took soundings around the nation and "encountered enthusiasm bordering on euphoria." Pollster Albert Sindlinger said, "In all the years I've been doing this business . . . I've never seen anything this unanimous unless maybe it was [the reaction to] Pearl Harbor." A poll by Opinion Research Corporation indicated, "On every specific action taken by the President, a majority of the public approved."

That same day, the *New York Times*, which had been consistently critical of Nixon's policies, ran an editorial stating, "We unhesitatingly applaud the boldness with which the President has moved on all economic fronts." *Time* wrote of Nixon, "His trip to China is almost certain to bring him political rewards, but come

Election Day 1972, mending the nation's pocketbook could pay off at the polls as Peking never would."

Many economists were laudatory, too. Walter Heller, chairman of President Kennedy's Council of Economic Advisers, proclaimed, "It's a historic initiative. The economic world will never be quite the same again." Robert Triffin, among the most esteemed academic authorities on global finance, wrote, "Finally the world has been forced to look the problem in the face, instead of trying to patch up the system."

The Democratic opposition had two reactions. They were grateful that the president had taken their ideas regarding wage and price controls, tax incentives, and dollar devaluation. But they did not like the idea that Nixon was getting all the credit. Wrote *Time*, "The Democrats have been embarrassed by this President who opened their closet and stole their shoes."

Some commentators raised fundamental questions, however. *BusinessWeek* drew attention to the implications of the government's unabashed intervention in the economy. "For a long period ahead, the U.S. [government] will be an unseen but very real presence at the bargaining tables where wages are set and in the board rooms where prices are made." Others worried about the broader international setting. The *Wall Street Journal* said, "The big question will be whether the world will accept what emerges [from the fallout of the NEP] as the basis for renewed building of trade and commerce." *Time* warned, "The hidden danger in the latest world monetary crisis is that if it is not resolved quickly and well, the world could tumble into a period of economic isolationism."

The wage and price freeze came under stronger attack than anything else. In his *Newsweek* column of August 30, Milton Friedman spoke for many free-market advocates in academia and business when he said, "Sooner rather than later, and the sooner the better, [wage and price controls] will end as all previous attempts to freeze prices and wages have from the time of the Roman Empire . . . to the present, in utter failure and the emergence into the open of the suppressed inflation." While most Democrats favored controls, some carped that the administration's plans were inequitable because they came down harder on workers than on business. They pointed to the plan's clamping down on all wages, even raises negotiated before the freeze and due to take effect in the next ninety days, while no lid was being placed on interest or dividend payments, which benefited mostly businesses and investors. While businesses could not raise prices, moreover, no limit was put on the profits they could earn, thus encouraging them to increase their margins by cutting expenses, including reducing their workforces. As for labor, although the unions supported a cheaper dollar and the import surcharge, leaders such as George Meany, president of the AFL-CIO, called the plan "Robin Hood in reverse, robbing the poor to pay the rich." Leonard Woodcock, president of the United Auto Workers, said he was ready to declare war on the administration, if that's what Nixon wanted.

NOT SURPRISINGLY, THE reaction of America's trading partners was one of extreme concern about possible financial chaos, mixed with a certain resignation that a wave of disorienting change was about to wash over the world. They were also angered about the abrupt, unilateral nature of the U.S. decision.

When Nixon's announcement was made, at 9:00 p.m. Eastern Daylight Time on Sunday, August 15, it was 10:00 a.m. on Monday, August 16, in Tokyo and still hours before sunrise on Monday in Europe. That morning, vacationing Japanese officials were called back to Tokyo to their head offices. When daylight came to the Continent, Canadian prime minister Pierre Trudeau abruptly ended his cruise off Yugoslavia to rush back to Ottawa, and French president Georges Pompidou hurried back to Paris from the French Riviera.

In Japan, the stock market dropped 20 percent during the first week. Japan kept its currency markets open, however, and it continued to buy dollars and sell yen in order to prevent the dollar from sinking and the yen from rising, with the goal of maintaining the 360-yen-to-the-dollar link that had been prevailing before the Camp David meeting. This frenetic buying lasted about two weeks and proved too costly, for Japan had to sell its government securities in order to buy about $3–$4 billion. In fact, it seemed that no amount of purchasing dollars could prevent the greenback from sinking against the yen, because global markets were forcing a de facto dollar devaluation by selling it in all markets.

In Western Europe, many markets were closed for a holiday on Monday, August 16. The following day, stock markets crashed everywhere on the Continent. Finance ministers and central bank officials decided to keep the foreign exchange markets closed for two weeks, during which time they hoped to sort out the wreckage. Within two weeks, both Japan and most of Western Europe allowed their currencies to float, but within limits. That is, they didn't take their hands off the rudder entirely, intervening continually to keep their currencies from floating upward too much and too fast against the dollar.

Outside the United States, concerns mounted about what specifically would happen next. The global framework for trade and investment had been frontally challenged. Big questions loomed abroad. Was the United States still committed to maintaining and strengthening a multilateral system of free trade and capital flows, or was it retreating with the goal of redefining its interest in a much more strident, nationalist way? Once an exchange rate alignment took place, did Washington intend to return to fixed exchange rates, where the dollar would be backed by gold, or did it have a totally different idea in mind—say, floating rates in which gold didn't play any role? The one reaction that all countries outside the United States shared was opposition to the import surcharge. Its coming from the world's leading evangelist of free trade was bad enough, but the indeterminate nature of it was even worse. The tariff was seen overseas as an odious American power play that could backfire and usher in an era of protectionism. Foreigners were also united in opposition to investment tax credits that would be available only to U.S. firms and not to foreign-owned operations within the United States. This was a discriminatory measure at odds with trade principles the United States had embraced for decades. These trade and investment provisions heightened anxiety by prompting memories of the economic warfare that characterized much of the 1930s still fresh in people's minds in European capitals and in Tokyo.

The two critical countries for the United States were Japan and West Germany, the two with the largest surpluses and strongest economies and the two with the most extensive military defense relationships. Tokyo and Bonn were in much different positions. Toyoo Gyohten, the junior official at the Japanese Ministry of Finance who scrambled to hear Nixon's speech on August 15, later described Japan's situation. "The Japanese were too naïve in believing President Johnson and President Nixon when they repeatedly pledged that the United States would not devalue the dollar," he wrote. He recalled how, at the Munich bankers' meeting just eleven weeks before the Camp David meeting, Secretary Connally had publicly repeated the pledge not to devalue. The Japanese could understand the closing of the gold window, but they never thought the United States would seek to break the fixed link of 360 yen to $1 that had been in effect for almost two decades. That they were so wrong showed how out of touch they were, Gyohten wrote, and why they were so shocked.

When it came to West Germany, communications between Washington and Bonn had been deeper and clearer. Japan had no close allies except Washington, while West Germany had the benefit of the other five members of the European Community plus the United Kingdom, making it better prepared for a big turn in the road because it could fall back on strong economic and political ties with its partners. Accordingly, Germany seemed more philosophical and resigned to certain realities that Japan didn't want to face, in part because it had already revalued its currency. A key editorial theme in German newspapers then was that U.S. leadership was in decline. On August 17, the *Süddeutsche Zeitung* said, "The dollar has collapsed as a leading currency . . . There is almost a declaration of an American trade war." The *Frankfurter Allgemeine Zeitung* wrote, "Nixon's program . . . documents a relapse of the world's strongest economic power into nationalism and protectionism." Kissinger later wrote, "The immediate significance of the new program was its effect abroad; it was seen by many as a declaration of economic war on the other industrial democracies, and a retreat by the United States from its previous commitment to an open international system. Clearly we were headed into a period of intense negotiations, conflict and confrontation."

How foreign officials and the public saw the entire package was not disconnected from how they viewed Secretary Connally. He was not a member of the financial club, and no one knew quite what to make of him. Finance ministers and central bank governors were "stunned and flabbergasted and appalled by his crafty methods," wrote British journalist Henry Brandon. "They were used to following rules of quiet dignity in their negotiations, of only lifting the corner of each of their cards and letting others just peek under them, but they were not used to having the whole pack thrown at them."

AT THE WHITE House, Monday, August 16, began with the president asking Haldeman to release a photo of his four top advisors—Connally, Burns, Shultz, and McCracken—from the previous day at Camp David. Nixon wanted to show that the decisions had been made with the wholehearted support of his economic team. That morning, he held a meeting to brief all cabinet members who had been excluded from Camp David. He emphasized that the New Economic Policy had been a long time in the making, that he had felt for years that Bretton Woods needed a new foundation, and that the United States had to take the lead in negotiating a new set of rules and obligations. He said the wage and price freeze would need a sequel of some kind, but he ruled out any kind of permanent controls. At the meeting, Burns was effusive in his praise of the president, saying that Nixon had electrified the nation. The Fed chair told the group that he had called dozens of central bank presidents and that their initial reaction had been "extraordinarily good."

That day also, Paul McCracken sent Nixon a memorandum reporting on the stock market, bond prices, housing starts (an economic indicator for the number of units of privately owned housing), and the "ebullient sentiment" of the day. Haldeman notes in his diary that Nixon was obsessed with domestic and international reactions. The president compared the New Economic Policy to the China opening and craved credit for being an extraordinary leader.

Later that morning, Connally briefed the press, with Burns at his side. It was a typical performance by the former governor—clear, confident, laced with humor, acknowledging that many questions were still unanswered, and with no trace of embarrassment or insecurity regarding what he didn't know. Asked to comment on Nixon's about-face on several policies such as wage and price controls, Connally replied, "There is a saying that there is nothing constant except change. The American people would think they have a dolt for a president if they had one that they thought would take a position and never change it."

ON MONDAY, AUGUST 16, the administration started rolling out its program. Having flown all night, Paul Volcker landed in London early in the morning. He proceeded to Wychwood House, home of the U.S. ambassador. Volcker was in Western Europe meeting with his counterparts from six countries, including Britain, France, West Germany, and Italy, who were joined by some more junior representatives from Japan who just happened to be vacationing on the Continent and had been ordered by their Tokyo bosses to show up at the briefings. Late that afternoon, Volcker and the ambassador, plus two others who were traveling with Volcker, were joined by twelve foreign officials.

Volcker saw his mission as beginning the process of reconstituting fixed exchange rates, albeit with a new realignment of values and with some more flexibility for exchange rates to fluctuate. He opened by saying he was not there to negotiate but, rather, to explain the central elements of Nixon's address of the night before. There would have to be negotiations, of course, and he said he would solicit views about where and in what forum they should take place. His demonstrative interest in listening rather than lecturing contrasted with the aggressively unilateral action that the United States had just announced.

Volcker explained that Nixon had decided not to act in a piecemeal fashion, but rather, to launch a comprehensive and integrated program. The United States had no long-term plan and was not about to spring one. Volcker did identify one goal that would guide U.S. policy: the United States needed its major trading partners to follow policies that would result in a major turnaround in the U.S. trade balance. That turnaround would have to not only erase the deficit, but also result in a small trade surplus. Volcker didn't want to be drawn into a discussion of whose exchange rates should move in order to cause such a major shift in the United States' deficits, but he did indicate that Washington had no intention of devaluing against gold in order to force the changes it needed. The dollar would continue to be valued at $35 an ounce. This, of course, implied that others—including, most important, West Germany and Japan—would have to revalue their currencies against the dollar. Volcker also mentioned that part of the solution to the U.S. trade deficits would have to be major trade liberalization in Western Europe and Japan and also a better sharing of the burdens for defense expenses. The message was that the onus for adjustment was on Western Europe and Japan, not on the United States.

Four concerns among the foreign officials were paramount. One: when would the "temporary" import surcharge end? Volcker said that the United States would have to be satisfied that policies would lead to a positive swing in the U.S. trade position. To that end, it was not enough to eliminate the merchandise trade deficit; a modest surplus had to be in sight as well. A second concern: when would the United States be willing once again to convert dollars into gold? Here Volcker punted and said Washington had no particular time frame, but that it did want to deemphasize the role of gold in the monetary arena, implying more use of the IMF's SDRs.

The third concern was this: Would the United States hold the dollar at current levels by making sure interest rates were attractive to dollar holders? In other words, would Washington push up interest rates so as not to allow the dollar to sink while talks were proceeding, thereby preventing a "sneak" devaluation? While Volcker gave no such commitment, he left little doubt in the room that Washington wanted the dollar to fall to lower levels. Like Volcker, most of the men in the room were proponents of fixed exchange rates, and a potentially long period of floating in which the dollar depreciated and many of their currencies appreciated made them nervous, if not angry.

A fourth concern was that the United States was counting on all the adjustments to fall on other countries and that when it came to the dollar and its trade competitiveness, Washington was making no concessions itself. Foreign officials did not consider wage and price restraints as a concession, but rather something the United States should have done in its own interest. In fact, a disconnect existed between the mind-sets of foreign governments and that in Washington. Most Western Europeans and the Japanese believed that the United States was substantially to blame for the current monetary disarray. In their view, the United States had mismanaged its economy by allowing inflation and unemployment to rise at the same time, and it had neglected to deal with its trade balance by failing to invest enough in technology and other areas that would boost its competitiveness.

Ultimately, nothing would be settled at this meeting, nor at a follow-up one in Paris with France's minister of economy and finance, Valéry Giscard d'Estaing. This was a much smaller gathering, with just a few officials present. The French minister reflected France's strong opposition to virtually all U.S. policies. The United States, he felt, coining an expression, had an "exorbitant privilege." His argument was that because the dollar was the predominant international currency, one that everyone could and did use for trade and investment, Washington could run huge balance-of-payments deficits and finance them by selling U.S. Treasury bonds, which investors around the world would be willing to buy. In other words, Washington could run deficits without the discipline that constrained all other governments. Giscard d'Estaing explained that the United States could therefore flood the world with dollars, many of which could be used to purchase European companies. In pushing this line of thought, he was reflecting the view of President de Gaulle, most of the French establishment, and many Europeans who were less willing to stand up to the United States.

Giscard d'Estaing argued for a onetime devaluation of the dollar, a continuation of fixed exchange rates, and the need to retain gold as an anchor of the international monetary arena. His position on gold was critical, for as long as the dollar was tied to the precious metal, Giscard d'Estaing believed there would be a modicum of discipline exerted on the United States. Nixon's bombshell announcement of the closing of the gold window was deeply unsettling. "France was concerned that everything that the U.S. had achieved over the past twenty-five years in terms of monetary stability and free convertibility would be lost," read the text of an American note taker for the meeting. "Mr. Volcker replied that we equally recognized the danger, but there was even a greater danger of growing protectionism in the United States if we had not acted to prevent further erosion of our [balance of payments]." In meeting with Giscard d'Estaing in so personal a way, Volcker was reflecting Washington's view that the French government was key to any monetary settlement. This was because France had been more opposed to U.S. policies in the past than other major European countries and also because the United States understood that, at the time, France had extraordinary political influence in the European Community.

LATE IN THE evening that Monday, August 16, Kissinger, back from Paris, spoke to the president by phone for about fifteen minutes. He found Nixon in a state of elation and feeling as if he had revolutionized economics just as he had transformed diplomacy. "We stirred them up a bit," the president said. Kissinger proceeded to lavish praise on his boss. "The thing that's so interesting about your style of leadership is that you never make little news . . . Mr. President, without you the country would be dead." Nevertheless, Kissinger sensed he would need to be involved in the follow-up to the Camp David decisions.

Indeed, his staff was on high alert. NSC staffer Robert Hormats had been following developments carefully and was being besieged both by colleagues at the State Department and by representatives of foreign governments to explain what was happening in the White House and Treasury. Realizing the huge foreign policy implications, he organized a conference call with two trusted and experienced advisors from outside the government who were confidants of Kissinger's. Their remit was to figure out what Kissinger should say to Nixon, Connally, and Burns. On the phone with Hormats was Richard Cooper from Yale, a prominent economist who had been in the Kennedy and Johnson administrations, and Francis Bator from Harvard, a lawyer and economist who had been a mainstay of LBJ's National Security Council staff. The phone call lasted a full seven hours. They went over every aspect of the August 15 decisions and the international economic and foreign policy implications. Among their major conclusions was the need for constructive multilateral negotiations, as opposed to the United States' bullying each country bilaterally. Also high up on their list of concerns was the danger the import surtax would pose to the alliances and the trading system itself if it were allowed to linger too long.

ON TUESDAY, AUGUST 17, President Nixon sent five planes to round up key lawmakers from their vacations for a briefing. He spoke to the bipartisan leadership of Congress together with the chairmen and ranking members of key congressional committees. At the meeting, he was particularly solicitous of Chairman Wilbur Mills of the House Ways and Means Committee, whose help he would need on passing tax legislation and other issues. At one point, the president nodded toward the congressman when explaining the key elements of the new plan. "We can all take credit for the program," he said. "Wilbur, these are some of your ideas."

About a week later, John Connally and Paul Volcker held a meeting with a small group of outside experts who had served in government and had considerable experience in international economic issues. (The exact date of this meeting may have been a few days earlier. Records are ambiguous.) It was one of several such meetings. Among those present were Fred Bergsten, who had been on Kissinger's NSC earlier in the administration, and two Yale economists, Richard Cooper and Henry Wallich. Connally began the meeting with a succinct question. "Okay," he said. "You know what we've done. Now what do we do?" Connally and Volcker mostly listened to the vibrant exchange among the experts. Although the discussion roamed far and wide, including issues relating to objectives, tactics, legal foundations for action, and long-term U.S. objectives, the group generally endorsed the administration's goals. Cooper and Bergsten, however, pushed hard for the administration to use this opportunity to press for substantial reform in the international monetary system. It wasn't enough for the United States to achieve a devaluation. Many other issues, such as the role of gold, and the expansion of IMF lending capability, should also be addressed, they said.

Still, Bergsten was unnerved by Connally's strident nationalist attitude, and he immediately walked over to the West Wing of the White House to see Kissinger and implored him to head off what he thought might be Connally's reckless approach to international negotiations. Kissinger was now being pushed by Hormats, Cooper, Bator, and Bergsten—all economists with public policy experience—to become centrally involved in the upcoming monetary talks.

ON THURSDAY, SEPTEMBER 9, President Nixon went before a joint session of Congress to present his new economic policies. He solicited support for the program in general and asked for passage of the tax-related provisions in particular. "We have fought two costly and grueling wars. We have undergone deep strains at home as we sought to reconcile our responsibilities abroad with our needs in America," he said. "In the years ahead, we will remain a good and generous nation, but the time has come to give a new attention to America's own interests here at home." He also reaffirmed his intention to keep the nation outward looking when he said, "We cannot remain a great nation if we build a permanent wall of tariffs and quotas around the United States . . . We cannot turn inward. We cannot drop out of competition with the rest of the world and remain a great nation."

In a staff meeting at the White House on Saturday, September 11, Nixon laid out some of his feelings about why America had to remain tough in its upcoming negotiations with Western Europe and Japan. The gist was that a strong nationalist policy was necessary to offset the isolationist impulse in the United States. "Now, if we give up too soon, we back down, we decide we are giving in . . . and [act as if] we are going to be responsible, and we are going to be good neighbors, and we are going to grin and bear it, believe me, the American people are going to say what the hell, we thought we had a president who was going to stand up for us," he said. "My point is, we are in a period where the United States, the people of this country, could well turn isolationist unless their president was looking after their interests. And we must not let this happen."

Later that week, Nixon held a White House news conference in which most of the questions focused on the China trip and the new economic policies. Reporters asked about the import surcharge and the specter of America's turning inward. Nixon said he anticipated long, tough negotiations and that the surcharge would not be coming off soon. The United States would not be isolationist, he said, so long as it was strong. "We have to have a strong America, strong economically and strong in the sense of its competitive spirit if the United States is to continue to play a vigorous activist role in the world," he said.

AT THE SAME time, Connally was in London to meet with other finance ministers and with central bank governors from Western Europe and Japan. Although Volcker had met with his counterparts in late August and again in early September, this gathering was the first meeting of finance ministers and central bank governors since the Camp David weekend. Those assembled were called the Group of Ten, or G-10, an assembly from the largest and most powerful countries in Western Europe, plus the United States, Canada, and Japan. The gathering took place in Lancaster House, near Buckingham Palace, in the grand music room with Corinthian columns, where Frédéric Chopin once serenaded Queen Victoria. The setting was apropos of the image surrounding the somewhat grand and mysterious club of high-finance officials who usually treated themselves to splendid surroundings that differentiated them from the ordinary work of civil servants.

The talks opened with considerable tensions. It was well known that the positions of Western Europe and Japan, on the

one hand, and the United States, on the other, were diametri-
cally opposed. What was also coming into focus were the broader
stakes involved. "The danger is that intransigence by the United
States could spill over into escalating trade reprisals and other
forms of economic retaliation," wrote Clyde Farnsworth of the
New York Times as the meeting began. "The long-range effects
of economic war would almost certainly be a worldwide recession
with uncertain political consequences."

Connally got right to the point. To achieve its goals, he said,
the United States would need a positive swing in its merchandise trade balance of $13 billion. Foreign officials were shocked at
the size of the figure. The treasury secretary acknowledged that
it might seem like a stunning shift, but said that it was based
on conservative assumptions about the underlying trends in the
U.S. trade picture, including the need to offset rising military
and foreign aid costs. The more bitter argument took place over
how to achieve what the United States wanted. Those who opposed Connally wanted assurance that the burden of realigned
exchange rates would not just be on them but also on the United
States; they demanded that Washington make a contribution to
the overall realignment by devaluing the dollar at the same time
that West Germany, Japan, and other governments revalued.
They also wanted Washington to quickly remove the import surtax, which was seen by America's trading partners as a powerful
weapon to pressure them not just to negotiate, but to capitulate
to all U.S. demands.

Connally gave no ground. His remarks were characterized
as "uncompromising, hardnosed." He said the United States had
done a massive amount for other countries over the past quarter
century and that now it was payback time. He demanded a redis-
tribution of monetary, trade, and defense burdens. Further, there
would be no contemplation of U.S. retreat on the import surtax

until his demands for exchange rate realignment and for a more equitable sharing of economic and defense burdens were acted upon. He went on to say that there should be no doubt that the United States had been a champion of open trade. "We believe in it, we fostered it, we spent much of the resources of our nation to support it and bring it about over the past twenty-five years . . ." And he pointed a finger at those countries whose governments had large trade surpluses and to the consequences of not reaching agreement. "But no nation should, over any period of time, assume that the export markets should be used for the purpose of providing prosperity at home to the detriment of other nations around the world . . . Let me conclude by simply saying that I suppose we of all countries would be most grieved to see a deterioration in the expansive world trade policies that have been built up in the last quarter century."

Connally infuriated other officials by refusing to propose a specific plan for what each country should do regarding its own exchange rates. He wouldn't say which governments should revalue their currencies or by how much. The officials at the meeting were used to Washington's providing concrete ideas, which was its leadership style. In this case, however, Connally was advancing just a target for the United States—the $13 billion reversal in its trade balance—and was saying that given that the United States would not devalue, it was up to the others to negotiate who should act and how. "We have not come here with any precise plans or details worked out," Connally told the group, "simply because it's obvious here that there are going to have to be, very frankly, considerable negotiations between countries [aside from the United States]." He was referring to the nations of the European Community, who would have to arrive at agreements among themselves in order to present a common position to the United States.

The media condemned the meeting as a failure. It reported that Connally had told delegates that he hadn't any intention of changing his position "by one iota." The event ended in deadlock, wrote William Keegan of the *Financial Times*, "with the Finance Ministers failing even to agree on the order in which they are to approach the problem of a currency realignment." A *New York Times* editorial said that if the United States remained so intransigent, "[i]t will almost certainly force foreign retaliation and a breakup of the free world's unity."

During a break in the conference, Connally commented to reporters: "We had a problem and we are sharing it with the world, just like we shared our prosperity. That's what friends are for."

A CABINET MEETING chaired by Nixon a week later was devoted almost entirely to international monetary negotiations. Connally explained that other countries were adamant about wanting an immediate removal of the import surcharge, and they wanted the exchange rate realignment to include a U.S. contribution of some decline of the dollar against gold. Connally intimated that he had no intention of caving. Nixon backed Connally, saying that the treasury secretary should not be concerned with the criticism he was getting from some quarters for not giving away the store. The president reminded everyone that the world of currencies had been undergoing one crisis after another, and that although the bankers wanted to return to the status quo, this would not be possible. Nixon also mused about the impact of these economic tensions on America's political and military alliances. In his diary, Haldeman quoted Nixon as saying that it was essential not to allow economic difficulties and differences to shatter the free-world alliance. According to Haldeman, though, the president indicated he didn't think it would get to that stage. Nixon gave a pep talk to those gathered around the table. "We must exercise leadership, because none of them will . . . We can only lead if we are strong economically, otherwise the American people will turn inward . . . We have to strengthen our own situation. Only then can an American president lead the world towards free trade . . ." Nixon talked about growing political and economic isolationism. "We can't just preach about it, we have to get at the heart of the problems, and that's the economic problem . . ." The question of how long to maintain the tariff surcharge came up, and almost the whole cabinet supported delaying any removal, as did the president. In his diary, Arthur Burns recalled that he was the lone objector, saying the surcharge was proving to be too confrontational. Burns was worried that the Western Europeans would soon retaliate and poison the atmosphere for any further progress on monetary questions. Indeed, Burns had been compiling a list of retaliatory measures that he heard other governments were preparing.

Both Nixon and Connally seemed confident that the United States was in a strong position to call the shots and that no big concessions would be necessary. "We have to remember that everyone wants to get into the U.S. market," Connally said of the U.S. trading partners. "We can't have a trade war, because they can't afford it." Haldeman recorded, "The President closed on the note that we should remember that a smile costs nothing, and that's all we'll give them."

STARTING ON THURSDAY, August 19, and continuing well into the fall, a number of congressional committees held hearings on the New Economic Policy. The Joint Economic Committee, comprising both senators and congressmen, was the group that had most consistently focused on the international monetary and balance-of-payments situation. It was first off the mark with hearings that began just four days after Nixon's announcement and continued on and off in thirteen different sessions until Thursday, September 23. The committee had broad jurisdiction that covered not just U.S. international economic policy, but also issues such as wage and price controls, and its oversight now dealt with the entire set of questions that had been raised by the New Economic Policy. The Senate Subcommittee on International Trade, a subcommittee of the Senate Committee on Finance, held hearings on September 13, 14, and 21 focused on the United States' competitive position in the world. The House Subcommittee on Foreign Economic Policy, a subcommittee of the House Foreign Affairs Committee, held hearings on September 16 and 21 concerned with the foreign policy dimensions of the New Economic Policy. All these congressional events were considered oversight hearings; that is, they were not designed to consider specific legislation but to raise the broad policy issues and allow experts from within the government and without to express their views. Their primary purpose was to educate senators and congressmen, as well as the public, and to lay the basis for more focused congressional examination later. The hearings involved an extraordinary range of economists and government officials, current and past. It was, by any measure, a textbook example of thorough and thoughtful congressional oversight.

Although both houses of Congress were led by the Democratic Party, the tone of the hearings was enthusiastically supportive of the August 15 package. General agreement existed that Nixon had taken exceptionally bold and necessary steps in a comprehensive and integrated package of domestic and international economic policies. Many of the challenges that would be raised in subsequent months and years were brought to the fore that fall: What comes next for wage and price controls? Is the domestic stimulus enough? Will the budget deficit spin out of control? What kind of exchange rate system should we have over the long run—fixed linkages with some modest flexibility, or freely floating rates? What should be the role of gold in international monetary affairs? How should global trade be made fairer and attack not just tariffs but non-tariff barriers, such as subsidies, quotas, and health and safety regulations that keep foreign products out? What does defense burden-sharing mean—paying more for U.S. troops, buying U.S. equipment, paying more for a country's own national defense, or all of these? How can the United States become more competitive in global markets?

At the Joint Economic Committee hearing on Thursday, August 19, Chairman Sen. William Proxmire (D-WI), no fan of the Nixon administration, set the tone for all the congressional hearings to come. "Today the committee opens one of the most important sets of hearings in its history," he said. "The President has just brought about a drastic change of course in public economic policy, one that switched the Nation from a passive economic policy to an activist one. The President has released what I refer to as an economic bombshell." Proxmire continued with a passionate soliloquy on the need for public understanding of what was at stake.

In opening the hearings of the House Subcommittee on Foreign Economic Policy a month later, on Thursday, September 16, Chairman John Culver (D-IA) said, "We have entered a complicated yet, nonetheless, decisive period of international adjustment to the new realities of the world economic scene. These matters have risen to the very top of the international agenda. How they are handled and resolved are questions that reach to the heart of our national interest as well as to the conditions of life for every American."

In the Senate Subcommittee on International Trade held on Monday, September 13, Sen. Abraham Ribicoff (D-CT) said, "It was quite obvious we had reached a turning point in our commercial relations with the rest of the world. Now . . . the key question is, 'Where do we go from here?'" At one point in the hearings, Senator Ribicoff was questioning Paul Volcker, the first witness. Volcker had stated that it was essential to take account of other nations' views in the ongoing negotiations. The senator shot back, "If you were interested in getting their thinking, apparently you were not very much interested in how they felt before the program was announced. I would gather from what you have said that as far as the United States is concerned we are playing this whole thing by ear. We don't know where we are going and, as the press indicates, neither do you nor Secretary Connally in going to the upcoming Ministers' meetings." Volcker replied, "I wouldn't accept either of those comments, Mr. Chairman."

In fact, both Senator Ribicoff and Undersecretary Volcker were correct, especially regarding the next steps. When it came to the context of the New Economic Policy and the goals of the United States, the Nixon administration, through the voice of the president and Secretary Connally, was outwardly espousing a tough line and seemingly unswerving goals. Within the administration itself, however, a battle was brewing between those who favored a return to fixed exchange rates after a major currency realignment and those who wanted to move to floating rates. Another struggle was between the foreign policy community, which felt that the United States' hardball tactics—especially the specter of a prolonged import surtax and the refusal to entertain any formal devaluation of the dollar—risked creating a breakdown in political and military alliances, on the one hand, and the Treasury, which felt that confrontational tactics were necessary, on the other. These internal battles began right after August 15, but they gathered force during September.

AT THE END of September, a clear example of disarray in the administration took place on the eve of the annual meeting of the IMF, an assembly of the world's top financial officials and bankers. This event was Connally's first major opportunity to express U.S. views since the deeply contentious mid-September ministerial meetings of the G-10 in London. Among many of the delegates, expectation was high that the treasury secretary's hard-line views would have softened somewhat since the London meetings. After all, the U.S. position had been subjected to withering criticism from abroad. Foreign officials hoped that the London meeting reflected Connally's opening negotiating position, and they were now eager to hear more from him. Most urgently, they wanted Washington to lift the import surtax, which they felt was a crude and overly aggressive tactic. They were seriously concerned, too, about damage to their exporters' access to the U.S. market, on which they had become so reliant.

Paul Volcker and his team had prepared a speech for Connally that reiterated the American position taken in London but also outlined conditions under which the import surcharge could be lifted. Prior to the IMF meeting, the draft was circulated among a few in the administration. One of the recipients, George Shultz, wasn't happy about the section on exchange rates. The fact was, the decisions of August 15 had not wrestled with precisely what would replace fixed exchange rates. Shultz knew that Volcker and Burns wanted a return to them, albeit with a realignment of currency relationships shaped by a revaluation of the West German mark and the Japanese yen, but to him that was no more than a pale version of the old Bretton Woods Agreement. Shultz hoped for a clean break from fixed rates and favored letting all currencies float against one another permanently. He was also convinced that Nixon wanted to move decisively in this direction, or would at least support it.

Shortly before the IMF meeting in Washington began, Shultz invited Connally over to his house. At Shultz's request, Milton Friedman joined them. The OMB director was determined to persuade Connally to change course. In Shultz's mind, now was a unique opportunity to create this new floating-rate regime. After all, in the wake of August 15, governments were allowing their currencies to temporarily float. Such floating was often called "dirty floating," because of extensive government cheating that consisted of buying and selling their currencies to control their value, or else putting controls on capital flowing in and out. Shultz hoped this was a chance to transition to a "clean" permanent float. In the relaxed and intimate setting of his home, he and Friedman tried to talk Connally into embracing the free-floating idea. Connally was noncommittal and asked Shultz to speak to Volcker.

Shultz contacted Volcker and presented a new draft that he had written for Connally to use that advocated freely floating currencies as the U.S. objective. On the eve of the speech, Volcker and Shultz sat in a suite at the Sheraton Park Hotel, where the IMF meetings were being held, and stayed up until the early morning hours arguing about their respective draft speeches and underlying approaches. Neither persuaded the other.

The two men decided to present their respective positions to Connally. The treasury secretary took Volcker's recommendation. It is doubtful he had a strong conviction on the substance of the approach. Instead, he relied heavily on Volcker's expertise and relationships around the world and trusted him on these issues more than anyone else in the administration. Years later, Volcker wrote, "In my own mind, the Shultz 'bombshell' would plainly not be negotiable." Volcker was no doubt referring to the fact that no other country, with the possible exceptions of West Germany and Canada, wanted floating rates as the permanent policy. They all yearned for stability and predictability and hoped that, after a realignment of currency values, a fixed system, with some more flexibility to move exchange rates than they were given at Bretton Woods, would hold.

EDWIN DALE JR. of the *New York Times* captured the mood at the IMF meeting that began the next day. He wrote, "Only one thing is reasonably clear: Just about everybody is very angry at the United States, or, at least has grave new doubts about it." A few days later, *New York Times* columnist Leonard Silk put it this way: "The language of diplomats runs to understatement and euphemism. Yet the worry is genuine here—and people are expressing it in the corridors and lounges—that if prompt action is not taken to end this crisis the world economy and the political and military alliances among nations will be seriously damaged."

In his address to the IMF delegates, Connally made one major concession. For the first time, he indicated the conditions under which the United States would remove the import surcharge. In order for everyone to see exactly what various currencies were worth, he told delegates that Washington wanted to see a clean float—that is, each country should allow its currency to be set at a free-market level, devoid of any government intervention or controls. Once that was done, everyone could see how the market was valuing currencies, and the new currency realignment could be set. Connally was of course banking on the likelihood that, with governments not propping up the greenback by buying it and by not simultaneously selling their own currencies, the dollar would depreciate dramatically. Connally also stated a second condition for removing the import surcharge: the United States would need to receive specific and immediate trade concessions—that is, commitments by foreign governments to open their markets to certain U.S. exports by a specified amount of tariff or quota reductions. If these two requirements were met, Connally said, the United States would lift the surtax. Writing for the *Financial Times*, Peter Jay, an economist and former British diplomat turned journalist, put it this way: "Mr. Connally left no doubt that the United States is in deadly earnest about correcting its balance-of-payments deficit, partly by exchange rate means but also, and equally important, by getting a crowbar into the unfair trading practices of Japan and the Common Market."

The speech fell flat with finance ministers and central bankers. They didn't want to abandon their "dirty" floats, and they didn't want to negotiate trade deals under such pressure. They thus demanded that the import surtax be lifted without preconditions. They still felt the turnaround the United States was seeking in its trade balance was absurdly large. And they were determined that part of the overall deal would have to be some dollar devaluation against gold.

The only issue agreed upon at the IMF meeting was the need to separate broad future reforms in trade and defense issues from an immediate exchange rate realignment, plus some tangible trade concessions right away. That long-term agenda included the role of gold and the amount of flexibility governments would have to allow their currencies to fluctuate around an agreed parity. In the trade arena, the agenda included negotiations to reduce non-tariff barriers that impeded trade, such as extensive national health and safety regulations. In defense, it included meaningful increases in defense spending. But all these broader and deeper issues would be put on the back burner while the G-10 ministers and central bankers tried to resolve issues over the dollar.

On the latter, the United States and its allies remained at an impasse.

The Finishing Line

The final decisions about the international monetary negotiations were hammered out within the administration during October and November. During those two months, a series of internal battles and compromises took place among the president and his advisors, especially John Connally, Henry Kissinger, Arthur Burns, George Shultz, Paul Volcker, and Peter Peterson. As always, President Nixon was the final arbiter.

The issues all involved resistance to Connally's bull-in-a-china-shop positions combined with his close-to-the-vest negotiating style. When it came to the Texan's tactics and strategy, for example, Peterson felt that Connally was trying to negotiate across too many fronts without knowing what the United States wanted, let alone what trade-offs should be made. In late August, Peterson told the president that Connally and Volcker did not have the capacity for broad thinking and that it was dangerous to entrust everything to them and their team at Treasury. Peterson wrote a series of memos that laid out the need for thorough review of monetary, trade, and defense goals and how they related to one another. He crystalized all the differences in positions that existed within the administration and posed issues that had yet to be resolved: How long should the import surcharge be in effect? In the end, would the United States be willing to devalue against gold, and if so, by how much? What is the time frame for meaningful concessions from the allies on monetary issues, trade issues, defense issues—and are these time frames the same for each issue? What must be done bilaterally to reinforce the multilateral strategy? In raising these fundamental questions, Peterson was trying to show Nixon that Connally and Volcker were moving too far and too fast without an agreed-upon administration strategy.

The second outstanding set of issues was what should replace the old Bretton Woods arrangements. Volcker and Burns favored a return to rules with realigned exchange rates and more flexibility around them. George Shultz and Paul McCracken wanted to use the moment to move decisively toward free-floating rates. Months before, Treasury had done a study that concluded that the balance-of-payments turnaround it wanted would require a 13–15 percent devaluation of the dollar against gold. Burns and Volcker supported such a onetime devaluation and then a return to fixed rates. This would mean raising the price of gold in dollar terms such that it took more than $35 to buy an ounce of gold, thus diminishing the gold value of each dollar. Volcker was willing to see gold phased out as a primary asset in the monetary arena, to be replaced by SDRs. Burns's position on gold was less clear, although like many central bankers, he was probably in favor of as little change as possible. Shultz and McCracken had a much different position. They thought the market should determine the value of the dollar now and forever and that gold should no longer play a part in the currency regime. Connally was focused on a broader package. While he was still siding with Volcker on exchange rates, he also had his eye on the importance of trade and defense agreements, which he felt were even more important to the president's political success at home than currency matters. Connally never strayed from his belief that most Americans had little interest in what happened in international finance, because the issues were so arcane. Connally also opposed devaluing the greenback because only Congress could change the price of gold, and he feared that lawmakers would attach too many protection-ist riders to any legislation making gold more expensive in dollar terms.

A third disagreement was building quickly over the strains on foreign policy that were developing as a result of the import surcharge. Now Henry Kissinger was being drawn into the fray.

ON MONDAY, SEPTEMBER 27, Peterson, still uneasy about Connally's handling of the negotiations, called Kissinger to ask whether he thought that Nixon had any idea of the foreign policy tensions that were arising as a result of Connally's tactics. Kissinger said the president was totally unaware of what was happening abroad. On Tuesday, September 28, Hormats wrote to Kissinger, "Forces within the U.S. government which will wish to squeeze every ounce of blood out of Europe and Japan regard-less of the political costs will vie with those who wish to lift the surcharge without the quid pro quos which hardliners will re-gard as acceptable." Although Kissinger had no fixed views about international monetary arrangements, he became concerned about growing fissures in the alliance, particularly at a time when Nixon was about to make two historic trips, one to Peking and another to Moscow in the following year. Both Arthur Burns and Peter Peterson shared Kissinger's views that Connally was pressing America's trading partners too hard. They thought that the surcharge could easily backfire, and Burns in particular warned Kissinger that Europeans were preparing retaliatory trade measures such as tight quotas on U.S. exports and that a protectionist spiral was not out of the question. The surcharge was beginning to hit Western European exports hard, too. West Germany feared a 30 percent drop in its sales to the United States if the surcharge were prolonged. The European Community estimated that it could lose three hundred thousand jobs. On top of that, Europe was entering a recession, and Kissinger worried that if the United States were blamed, the battle lines would be hardened further.

In fact, members of the European Community were trying to coalesce into common opposition to the U.S. hard-line policy. Washington did not want its pressure in the currency negotiations to help push Western Europe into a bloc dedicated to opposing U.S. policies across the board. Also on Kissinger's mind was the sensitivity of Japan. The country relied almost entirely on the United States for its foreign markets and its military security. In addition to having been shocked and embarrassed by the announcement that Nixon was planning to go to Communist China, the unilateral August 15 measures constituted another massive surprise. Tokyo was also under excruciating pressure to reach agreement with Washington on quotas for its textile exports. Kissinger's anxiety was rising about the negotiations in general. "I came to the view that some shock had probably been needed to bring about serious negotiations," he recalled years later. "My major concern was to end the confrontation when it had served its purpose and to prevent economic issues from overwhelming all considerations of foreign policy."

Like the European allies, Kissinger did not approve of Connally's having set an overall goal—a $13 billion turnaround in the U.S. trade position—without specifying what each government should do to achieve that end. Kissinger preferred that the United States offer a detailed plan of its own, as consistent with the history of American leadership. While he was careful not to criticize Connally, he tried to make Nixon aware of the escalating tensions abroad. "I have to meet more regularly with Connally," he told the president on Thursday, September 30, "because Texans don't really have the diplomatic touch." Kissinger sensitized the president another way, too. He sought Connally out and tried diplomatically to make him appreciate the foreign policy stakes, then reported his efforts to Nixon. On Monday, November 15, for example, Kissinger recounted to the president, "I went over [to see Connally] and I told him I was not coming as the president's emissary, I'm coming as your friend." Kissinger continued summarizing what he had told Connally: "You've now smashed the system. No one would have had the guts to do it except you . . . Now I figure it's my judgment, if you want to emerge as a statesman out of this, you have to move into the constructive phase."

IN THE MIDST of the various conflicts and tensions among his advisors, Nixon kept everyone off balance. In one meeting, for example, he sided with Peterson's recommendation that Connally's solo act be subject to interagency guidance and restraint. At another time, he told Kissinger to make sure that the foreign policy dimension was part of the discussion. Then, fearing that Kissinger and Burns had gone soft on the allies, he urged Shultz to buck them up and not allow them to give away the farm. On occasion, he favored Connally's policy of prolonging the import surcharge, but later he agreed with Burns that Connally was taking the surcharge too far. Nixon would rail against the allies as needing a good thrashing. Then he would turn around and talk about the need to protect political and security relationships with

Western Europe and Japan. The one constant was that he always made it clear that Connally was in charge. For Nixon, the calculus was simple: Connally was the only one on his economic team who shared his broad political instincts and who had his eyes firmly on Nixon's reelection in 1972.

Maybe this vacillation was not surprising. After all, Nixon had to balance several considerations—the economy, foreign policy, the international monetary and trade negotiations, and considerations concerning his 1972 reelection. It was essential to him that the economy revive, employment go up, inflation go down, the balance of payments show progress, and his trips to Peking and Moscow be seen as grand triumphs. It was also crucial to him that he maintain the aura of the aftermath of the Camp David meeting and the glow of being perceived as a strong leader at home as well as abroad. On any given day, one priority might overtake another, but then, no one else had such a broad range of concerns and responsibilities.

BUT AS OCTOBER bled into November, Connally's tactics were beginning to worry Nixon. The president was hearing not just from Kissinger, Burns, Peterson, and Shultz, but also from friends on Wall Street anxious about rising economic tensions abroad. Peterson told Nixon that he was receiving many calls from business leaders who felt the administration didn't understand the precarious state of the stock market and other uncertainties plaguing the economy, and that he had been asked to tell the president and Connally to "cool it," quit the "saber rattling," and "stop the don't-give-a-damn" attitude when it came to monetary and trade negotiations. Nixon was also receiving cables from his ambassador to France, Arthur Watson, and his ambassador to West Germany, Kenneth Rush, that U.S. pressure was being deeply resented at the highest political levels in their respective countries.

On Monday, November 1, Hormats advised Kissinger to persuade Shultz, as someone whom the president deeply trusted and who was respected by all parties, that it was in the country's best interest to conclude the negotiations soon, before matters got out of hand. This meant ending the surcharge in return for revaluations of other currencies, some small trade concessions, and some modest offers from West Germany and Japan to pay more for defense. In fact, Shultz shared this view. He was a master of accepting half a solution if he thought the rest would be possible at a later stage.

The following week, Kissinger described the state of negotiations to Nixon. "We are uniting all those countries against us by not telling them what we want . . . If we screw everybody in the free world . . . we will then undermine the whole structure of free world competition." Kissinger was particularly worried about Connally's repeated public statements that the surcharge would remain in effect for a long time.

THE PIECES OF an endgame fell into place, particularly during meetings in the Oval Office on November 22, 23, and 24, when intense discussions were held between Nixon, Connally, Kissinger, and Shultz. Looking toward December, Nixon had planned to see selected heads of state to brief them on his upcoming trips to Peking and Moscow. However, on Tuesday, November 23, in an Oval Office meeting to discuss the upcoming heads-of-state summits, Kissinger told the president that unless the international economic issues were settled beforehand, these sessions would be too fractious. (One person I interviewed believed that Kissinger or one of his top staff members may have encouraged European leaders to take this hard line, knowing that their concerns would register with Nixon and provide a route to bringing the divisive economic negotiations to an end. However, I could not verify this assertion.)

Connally was present and sensed the political winds. He knew how much Nixon's upcoming meetings in Peking and Moscow meant to him, so he preempted a potentially contentious discussion and proposed a way forward. He explained that the next meeting of finance ministers would be in Rome later in November and that he could conclude a deal only if the United States agreed to devalue the dollar against gold. The final result wouldn't be everything that he or Nixon wanted when it came to trade and defense, Connally admitted, but the United States could bring the negotiations to an end for now, leaving a host of other issues to be handled in a subsequent phase of negotiations.

Nixon jumped at the idea. "Make a deal on monetary things," the president directed. "Say the deal on trade will come later." They agreed that Connally would start the negotiations in Rome and Nixon would quickly follow up with his first summit, which would be with French president Georges Pompidou. Pompidou was a former banker, knew all the monetary and trade issues, and wielded more influence in the European Community than any other head of state. Nixon would then conclude a conceptual deal with Pompidou, and then Connally could fill in the last details with his counterparts at a subsequent meeting. "I discuss something with Pompidou, leaving something that he gets," Nixon said, thinking aloud in front of Connally and Kissinger. "He's got to have something to be credible . . . And then we turn back to John and you [John] work the damn deal out." It could all be done before 1971 was over. That became the plan.

On November 25, referring to the upcoming meeting of finance ministers and central bankers in Rome, the *New York*

Times ran an article, "Nixon Is Hopeful on Money Talks." The schedule of summit meetings, beginning with Pompidou in mid-December, had already been made public. The tide was turning.

A THIRD SET of ministerial meetings of the G-10—the first having been in London in mid-September, the second during the late September–early October IMF meetings—took place in Rome between Monday, November 29, and Wednesday, December 1. On the way to Italy, Connally confided to Volcker that Nixon had encouraged him to make a deal and that he was authorized to offer a modest devaluation of the dollar to seal it. Connally explained that Nixon wanted to do the actual deal with Pompidou at a presidential summit in December; therefore, the treasury secretary would have to open the door to a devaluation, smoke out others' reactions, but stop short of an actual agreement. Just before the meeting, Connally and Volcker discussed how that strategy could be orchestrated, devising a set of talking points each would use.

The Rome meeting was held at the Palazzo Corsini, an elegant structure built between 1730 and 1740 as an extension of a fifteenth-century villa, not far from the Tiber River. Meetings took place at a large, ornate rectangular table in a room with a marble floor, walls of Italian art, and a high, vaulted ceiling. Hopes were palpable that this meeting would make more progress than the previous G-10 gatherings. Paul Lewis of the *Financial Times* wrote that the indications were "that President Nixon is now personally anxious to resolve the world's monetary crisis sparked by his August 15 measures." An editorial in the same paper captured another point: "Mr. Nixon, who cannot hope to present himself in Moscow and Peking as the spokesman for the West as a whole while an economic squabble is dividing the alliance, will no doubt be seeking means of tactfully pressing the point home," it noted. "The years are gone when America could more or less insist on having its way, but the time has not yet arrived when Europe can manage without the United States. There is a need for give on both sides."

The official meeting began at 3:30 p.m. on Monday, November 29. The G-10 had rotating chairmanships, and for this particular meeting, it was Connally's turn to chair. This meant that Paul Volcker became the official representative of the United States. The meeting got off to a slow start. Italian and West German representatives made long, dull presentations that recounted all that had transpired since the Camp David weekend. At one point, Connally briefly fell asleep. The next morning, Volcker gave a summary of what the United States was demanding. He changed the target for the magnitude of the necessary reversal in the U.S. trade balance from $13 billion to $9 billion. That was not enough, Volcker said, but the United States was reluctantly willing to live with that smaller shortfall if others stepped up to serious negotiations to revalue their currencies. If U.S. requirements were met, said Volcker, then Washington would lift the surcharge. He referred also to the need for some concessions to the United States on trade and defense. None of the foreign officials was moved with what they saw as a restatement of the old U.S. position.

Connally, as chair, then suggested that they break and reconvene the next day. That afternoon, he and Volcker refined their plan for how they would break the impasse. When the session resumed on Wednesday, December 1, French minister Valéry Giscard d'Estaing said that unless the United States were willing to make some "contribution"—by which he meant, Washington would have to agree to some devaluation of the dollar against gold—there was nothing to discuss. Giscard d'Estaing was role-playing. He had been well briefed by his own foreign ministry and understood that the idea of a dollar devaluation couldn't be taken too far now so that the possibility of a dramatic break-through could be reserved for the upcoming Nixon-Pompidou summit.

Connally leaned over to Volcker and whispered something. It was evident that some signal had been passed between them. Volcker then addressed the group. "Well, suppose, just hypotheti-cally, we were willing to discuss the price of gold," he said, mean-ing a U.S. devaluation against gold. "How would you respond if we increased the price by 10 or 15 percent?" (This meant: how would they feel if the United States devalued against gold by 10 to 15 percent.)

Connally immediately intervened. "All right, the issue has been raised," he said. "Let's assume 10 percent. What would you people do?"

It was a historic moment, the first time that dollar devaluation was officially raised since the rates were set at Bretton Woods. Total silence followed. Volcker recalled that for the next hour, "There was no answer, no discussion, no attempt to change the subject, no call for recess . . . there was just silence. Some smoked, some whispered a little to their colleagues, some just fidgeted." As it turned out, none of the ministers had anticipated the United States would make this kind of concession, and so none had in-structions from their home government as to how to respond.

Finally, Karl Schiller, West German minister of economic affairs and finance, spoke. He said that if the United States deval-ued against gold by 10 percent, West Germany could revalue by that amount, plus 2 percent more—meaning a revaluation of the mark of 12 percent. After a pause, Anthony Barber, the British chancellor of the exchequer (the equivalent of finance minister), said he could never agree to a U.S. devaluation of 10 percent.

In his eyes, that would make America *too* competitive. Connally asked, "What would work?" Barber replied 5 percent. Connally said that would be too low to contribute to the major balance-of-payments turnaround the United States needed. The discussion ended there, inconclusively but just as Connally and Giscard d'Estaing wanted. The deal was ready for Nixon and Pompidou to finish off.

By the end of the meeting, the psychological barrier of a new gold value for the dollar, a devaluation, had been crossed. As the session was ending, the European Community had committed to entering long-term, broad trade negotiations, and it was agreed that defense issues would be handled within the NATO forum. Issues of longer-term monetary reform were put off for a second stage of negotiations—as had been agreed at the recent IMF meeting—which everyone understood would take a long time. Right now, the focus would be a currency realignment, some quick trade concessions, and some small gestures toward better sharing in defense spending. In the words of journalist Henry Brandon, this was the "moment of the formal dethronement of the Almighty Dollar." That turned out to be an overstatement, but it was the start of the first formal multilateral negotiations since Bretton Woods of a major realignment of key currencies.

The G-10 agreed to meet again in Washington on Friday and Saturday, December 17–18. By then, Nixon would have met with Pompidou, and hopes were high everywhere that a breakthrough was at hand.

THE NIXON-POMPIDOU SUMMIT took place on Monday and Tuesday, December 13–14, on Terceira Island, in the Portuguese-owned Azores, an archipelago of nine volcanic islands in the middle of the North Atlantic. During the following thirty-six hours, Pompidou did not want his finance minister, Valéry Giscard d'Estaing, to participate in several of the bilateral meetings with Nixon. Not only did the two men not get along personally, but Giscard d'Estaing was a potential political rival to Pompidou. Because Connally was Giscard d'Estaing's counterpart, for reasons of protocol, he, too, couldn't participate in many of the meetings. Such protocols didn't pertain to Nixon's national security advisor, and so Kissinger was catapulted into the center of the negotiations, taking the place Connally would have had. "So it happened that a solution to the monetary crisis was being negotiated between Pompidou, a leading financial expert and a [former] professional banker, and a neophyte," Kissinger wrote in his memoirs. "Even in my most megalomaniac moments, I did not believe I would be remembered for my contribution to the reform of the international monetary system."

It turned out that Kissinger, in close cooperation with Connally, rose to the occasion. On Monday morning, Connally briefed Kissinger on the key issues and on the minimal position that would be satisfactory to the United States. Kissinger then met alone with Pompidou, along with their interpreters. Kissinger told the French president he would like to know France's minimal requirements and that he, Kissinger, would take them back to President Nixon and then return with a response. Pompidou obliged, and Kissinger conferred with Nixon, Connally, and Volcker about next steps. After lunch, Kissinger, armed with detailed advice, returned to offer a compromise position to Pompidou. They were still far apart.

That evening, Monday, December 13, a banquet was hosted by the Portuguese prime minister. Nixon sat to the right of Pompidou, Connally to the left. In the discussions between Pompidou and Kissinger that day, the most critical difference was not whether the United States would devalue the dollar—that had more or less been decided at Rome—but the magnitude of that devaluation. At dinner, Connally and Pompidou held a quiet negotiation that moved toward agreement. If the United States was going to devalue, Connally wanted the biggest decline in the value of the dollar he could get, because the lower the dollar rate, the more competitive the United States would be. Pompidou began by saying the United States should devalue by 6 percent. Connally countered with 9.5 percent. Pompidou came back with 7. Connally offered 9. They left it there.

The next morning, Pompidou and Kissinger sealed the deal at breakfast. The United States would devalue the dollar against gold by 8.7 percent.

It is not clear how focused Nixon was on the monetary negotiations, as his interest in all these summits was solely on the grand political strategy. Kissinger later wrote that Nixon was not on top of his game. "If given a truth serum," wrote Kissinger, "he would no doubt have revealed that he could not care less where in the new scale the various currencies would be pegged." To be sure, Nixon was tired. While he was in the Azores, he had been focused on preventing a war between India and Pakistan, with all the around-the-clock cables, phone calls, and face-to-face discussions with Kissinger that task entailed. And after the banquet with Pompidou, he had decided to stay up most of the night listening to the game between the Washington Redskins and the Los Angeles Rams on the Armed Forces Radio network. At 4:20 a.m. Azores time, which was 12:20 a.m. in Washington, he awakened Haldeman, radiating excitement about the Redskins' victory.

But despite Nixon's lack of focus, the private agreement between the two presidents contained significant detail. The United States would devalue the dollar against gold by 8.7 percent, which meant that one ounce of gold would no longer equal $35 but would be worth $38. Other countries would be obliged to revalue their currencies by specified amounts, while still others would devalue. France would neither devalue nor revalue in terms of gold.

Once the currency realignment took place, currencies would be allowed more flexibility to move up and down. The Nixon administration promised to vigorously defend the new dollar exchange rate, making particular effort to control inflation with wage and price restraints and to increase productivity through investments in R&D. Ominously for the enforcement of the agreement, nothing was said about U.S. monetary policy and interest rates, which were the United States' most effective economic levers when it came to exchange rates. Of great significance to U.S. trading partners, however, the United States would remove the import surcharge and those parts of the investment tax incentives that discriminated against foreign companies.

This agreement implicitly acknowledged that, for the time being, the U.S. gold window would remain closed and that the replacement for gold would be defined and established in future negotiations on monetary reform. Once again, it was agreed that the European Community would immediately focus on getting a major round of trade negotiations off the ground. Reference was made to a recent NATO meeting that "represents a constructive approach toward dealing more adequately with a proper sharing of the defense burden."

As part of the closing ceremonies, the officials announced that the G-10 would be reconvening in Washington that weekend, beginning Friday, December 17, to continue detailed discussions.

When the delegation landed at Andrews Air Force Base late Tuesday afternoon, Nixon asked Connally to say a few words to the assembled reporters. "I think it is fair to say that the meeting between the two Presidents resulted in a very significant step forward," said Connally. "The Group of Ten meeting which will occur this weekend hopefully will bring us even closer to a solution." Hobart Rowen of the *Washington Post* was ecstatic. "The Azores agreement . . . is a momentous event in monetary history," he wrote, pointing out that the logjam blocking not just immediate currency realignment but also negotiations on fundamental reform of international finance and trade had now been removed. An editorial in the *Washington Post* captured the event in its largest sense: "The Azores communiqué is the economic counterpart of the Guam doctrine [i.e., the Nixon Doctrine]," it said. "It serves notice on the rest of the world that the United States is now defining its national interests much more tightly . . . Domestic considerations, above all high and stable employment, now count more in relation to our foreign responsibilities. The devaluation is a rational and useful adjustment, of a technical nature, to the mood now ascendant in the country."

In a morning meeting the next day, Nixon lectured Connally and Kissinger on the public relations imperatives. He wanted to be sure that it was known that he, Nixon, had hit hard the three issues of exchange rates, trade, and defense costs. Nixon wanted the press to reflect his intense determination to generate more jobs, too. When it came to the devaluation, he wanted it portrayed in terms of the United States' now being more competitive, rather than of any diminution of U.S. power or standing in the world. Don't say the dollar was devalued, he implored, but that it had been overvalued. In his diary, Haldeman recorded that either the president or Connally—he couldn't recall which—had talked about the need to be clear what the last few days were all about. "There should be an in-depth piece [in the media] that no one ever questioned the U.S. political and military leadership of the free world," one of the men said. "This move now results in U.S. leadership in the monetary and financial field, as well. We should compare the Azores to Guam and say that this is burden-sharing in the broadest sense." Nixon, Connally, and Kissinger went over the line of argument once again. They had been successful in realigning currencies, they said. They had forced the Western Europeans to focus on trade. They would make progress in NATO by increasing allies' percentage of military expenses. They kept rehearsing the arguments, convincing themselves that the Camp David decisions were indeed proving to be a major success.

AS SOON AS he returned from the Azores, Connally set out to find a suitable venue for the G-10 meeting scheduled for that weekend. He selected the Smithsonian Institution on the National Mall, a building he felt would set the right tone. Built just before the Civil War, it was as close as the United States had to a genuine castle, a red sandstone structure replete with towers, rounded arches, forty-four-foot-high ceilings, and tall, narrow windows.

Nixon called key senators and congressmen to seek their backing for a devaluation of the dollar relative to gold. He received considerable support as well as a commitment for quick congressional action to approve a change in the price of gold once he submitted legislation. The reaction from the business community to an upcoming exchange rate announcement was positive, too. CEOs cited the importance of ending the uncertainty surrounding the value of exchange rates as a key boost to new investment and expanded trade. The hope was that foreign investors would be coming into the United States to purchase assets priced in devalued dollars or to invest in land and factories that would be cheaper in terms of their revalued currencies. Thus, the CEOs also felt that the share prices of many U.S. companies would go up, propelling the stock market generally.

On Thursday, December 16, the finance ministers and central bank heads of the G-10 began arriving in Washington. As the meeting started the next day, the dollar began to sag in global markets, showing that banks and investors were anticipating a dollar devaluation. Between the Rome and Smithsonian meetings, for example, the dollar was down against the West German mark by over 12.5 percent and against the Japanese yen by nearly the same amount.

Connally, still the rotating chairman of the G-10, led all the sessions. As he entered the building on Friday morning, December 17, he told reporters, "We are prepared to push for a conclusion," and then he added, "I am not at all sure we can or not, and I don't want to be too optimistic that we can." Indeed, during the first day of the talks, little progress was made. Four difficult issues confronted the delegates.

The first was just how much the dollar would be devalued. The answer entailed not only how much lower the dollar would be in terms of gold, but how much higher other currencies would go in terms of the dollar—for it would be the spread between the dollar devaluation and the revaluation of other currencies that counted. At the Azores meeting, the United States had agreed to devalue the dollar against gold by 8.7 percent and that there would be no change in the value of the French franc versus the dollar. (In other words, the French would neither devalue nor revalue against gold.) But the two presidents were in no position to negotiate what the revaluation of the West German mark or the Japanese yen would be, let alone dozens of other currencies. Thus, a more comprehensive agreement on currencies was still necessary.

The second big issue was trade. All along, the United States had insisted that a good deal of its balance-of-payments deficits was due to protectionism on the part of the European Community and Japan. However, it had become clear since the Rome meeting that fundamental trade reforms would take much more time than an exchange rate realignment. The issue in the Smithsonian talks for the Americans was how to get the dollar down and how to achieve some small trade concessions quickly while locking in a firm commitment from Europe and Japan to proceed with negotiating more extensive trade concessions in the future.

The third issue was one that had not received too much attention since August 15. It was clear to all the officials gathered at the Smithsonian that the days of the United States pledging to convert excess dollars abroad in central banks into gold was over. The United States simply didn't have enough of the metal. But foreign governments wanted to be able to convert their excess dollars into some kind of reserves. What would the United States offer, what combination of some gold, some foreign currencies, some of the IMF's SDRs? There had to be some commitment from Washington, the foreign delegates said, or else what would stop the United States from continuing to bleed dollars, forcing foreign governments to hold more and more of them? Washington didn't doubt the need for convertibility of dollars to other assets of some kind; it was just demanding that this issue be considered after an exchange rate realignment took place and in the context of long-term monetary reforms.

A fourth issue was one of good faith—or lack thereof. In order to devalue the dollar against gold, it was necessary for Congress to pass an authorizing law. Foreign delegates wanted an assurance that this would happen. They were afraid that for domestic political reasons, the administration would postpone submitting legislation too long. At the same time, Connally wanted ironclad assurance that the Western Europeans and Japanese would move on trade issues before he asked Congress to act on gold. He therefore wanted to make submission of the legislation to Congress for the dollar devaluation contingent on trade negotiations taking place. So, the attendees were presented with a chicken-or-egg dilemma.

THE FIRST DAY of negotiations, Friday, December 17, ended badly. "Everyone was too concerned with protecting his own position, everyone's own nationalism was shining through," wrote Henry Brandon. The currency realignment itself was complicated; it was a question not just of how each country would fare competitively against the United States, but also of how they would fare against one another. So, for example, the French wanted a maximum West German revaluation against the franc so that French products would be cheaper in West Germany. West Germany, of course, had just the opposite interest. The Smithsonian negotiations were becoming three-dimensional. Connally recalled to writer Martin Mayer that much of the time was spent largely in recess, as finance ministers consulted with their home governments and as the six EC members tried to come up with a common position.

It wasn't until the second day, Saturday, December 18, that the negotiations began to take off. The Western Europeans could not agree among themselves, and the West German minister of economic affairs and finance, Karl Schiller, and the French minister of the economy and finance, Valéry Giscard d'Estaing, both came to Connally and said that the Americans would have to strong-arm each European delegation, one by one. It was, after all, the first time since Bretton Woods that the countries were negotiating not just their parities with the dollar but with one another. Connally was now in his element, horse trading. "He cajoled, threatened and roughed up these ministers," said Brandon. "He may not [have been] a monetary wizard, nor a financial strategist, but he proved to be a negotiating tactician par excellence who knew how to squeeze the last drop of blood out of those anemic keepers of their countries' treasuries."

First, Connally couldn't get the Japanese to budge. Then he approached West Germany's Schiller and proposed that his country revalue the mark by 14 percent. Schiller asked what the French would do. Connally said they wouldn't revalue or devalue, as had been agreed in the Nixon-Pompidou Azores summit. What about the Italians? Schiller asked. What about the Japanese? Connally assured him that he would demand that the Japanese revaluation be at least three percentage points higher than West Germany's. The treasury secretary went from delegation to delegation, calmly but forcefully negotiating deal after deal, making it clear that no deal was possible without their providing the devaluation the United States needed.

Connally met the Japanese delegation in a small conference room. He started out demanding that Tokyo revalue by 19.2 percent. The Japanese resisted. Connally retreated to 18 percent and then, still unable to move them, went to 17 percent. The Japanese still refused to budge. Finally, a Japanese official explained that in 1930, one of their finance ministers had revalued the yen by 17 percent. It was thought that this move started a recession in Japan and resulted in the minister's being assassinated. Connally paused a second. He understood the power of superstition. How about 16.9 percent? he asked. The deal was done. An overall agreement was in hand.

THE SMITHSONIAN AGREEMENT was seen as the first stage of a major overhaul of the Bretton Woods Agreement. Among the official delegates, including the Americans, no pretense was made that it constituted the full set of reforms that would be necessary; the intention was to embark on future negotiations for that. Likewise, the fierce determination of Nixon, Connally, and the other U.S. officials regarding major trade liberalization would be nowhere near satisfied by the concessions that the Western Europeans had agreed to discuss immediately; there, too, negotiations would be necessary down the road. The U.S. delegation knew that the extent of the currency realignment that they had agreed to in the Azores and again at the Smithsonian was less than the United States needed to turn around the growing trade deficit. That said, almost all the attendees at the Smithsonian felt they had responded as best they could to the August 15 shock and that a historic deal had been struck.

They had agreed to an exchange rate realignment. The dollar would be devalued against gold; now one ounce of gold would require $38 and not $35. However, the calculation of the overall outcome for the dollar was more complex, owing to the many currencies involved and the differences in the size of the various import markets. Even though Japan and West Germany had agreed to large revaluations in relation to gold, other countries such as Great Britain and France—also large markets for the United States—had decided not to devalue or revalue against gold. And still others, such as Italy and Sweden, had actually devalued. A large part of the world also continued to fix their currencies to the dollar, undercutting whatever advantage a devalued greenback would have provided. Canada, among America's largest markets, refused to fix its exchange rates at all and just let the Canadian dollar float against all other currencies. Experts talked of "trade-weighted" exchange rates to express the true impact of exchange rates on trade. Thus, when the smoke of arcane currency calculations cleared, the overall U.S. devaluation was a little shy of 8 percent when it came to all countries and about 10 percent when it came to the biggest and richest ones. To put it in more tangible terms, after August 15, 1971, when Volcker and Connally began bargaining with the rest of the G-10, they said the United States was adamant about a currency realignment that would result in a positive swing of the U.S. trade deficit of $13 billion. That sum would move the United States from a deficit to a slight surplus, given the negative underlying trade trends that the United States was projecting, and the increased costs of overseas military expenses and foreign aid. As a result of the Smithsonian Agreement, however, the Federal Reserve estimated that the trade balance would improve by $8 billion from its 1972 level. This would allow for no surplus at all.

Once the realignment was in effect, currencies would have more leeway to fluctuate around the new fixed parities before having to officially devalue or revalue. That leeway had been 1 percent under the original Bretton Woods Agreement, and now it would be 2.25 percent up or down from the fixed rate, amounting to 4.5 percent overall. For the foreseeable future, the dollar would no longer be convertible into gold or anything else, with the exception of some minor technical IMF transactions. This issue of converting dollars for gold would be dealt with in later monetary negotiations.

The United States had agreed to drop the import surcharge. It would also do away with those parts of the investment tax provisions that limited benefits only to U.S. companies—the so-called "Buy America" provisions.

There remained a critical question of how currencies would retain their new fixed value. What rules were they supposed to follow? What would be the enforcement mechanism? French president Pompidou had made it clear to President Nixon in the Azores that the new arrangements demanded that the United States get its own economic house in order, particularly when it came to controlling inflation, and Nixon had agreed. Would he deliver? Would the wage and price controls, now in a second and very controversial phase, work? Would Arthur Burns raise interest rates to keep the economy from overheating, or would he buckle to Nixon's pressure to lower rates so that the economy and employment grew in time for the U.S. presidential election, eleven months from then? How long would the broader monetary and trade negotiations take? The announcements of August 15, 1971, had an effect of moving the Bretton Woods system into a new era, but many fundamental questions remained unresolved. For now, though, almost everyone was satisfied with the results at hand. Among other factors, they were no doubt too exhausted to do more.

CONNALLY HAD RECEIVED word from Haldeman that if the negotiations succeeded, Nixon wanted to appear. As late as 12:45 p.m., Connally called Haldeman and said he was doubtful they would finish and that the G-10 may have to meet again the following month. Then the negotiating tide suddenly turned. Connally called Haldeman again. A deal had been struck after all, he said.

Nixon arrived and spoke to the delegates at 5:37 p.m. The group had all moved to the National Air and Space Museum, part of the Smithsonian. The president stood in front of the airplane the Wright brothers flew in 1903. "It is my great privilege to announce on behalf of the Finance Ministers and other representatives of the ten countries involved, the conclusion of the most significant monetary agreement in the history of the world," he said. It is a statement often quoted and ridiculed for its hyperbole. However, Paul Volcker put it in perspective years later. "The remark has been repeated with a scornful laugh," he wrote, "but as far as I know, it was in fact unprecedented to have so many countries agree on a set of exchange rates at one time."

Nixon went on to give the historical context. "When we compare this agreement to Bretton Woods . . . we can see how enormous this achievement has been," he explained. "Bretton Woods came at a time when the United States, immediately after World War II, was predominant in economic affairs in the world . . . Now we have a world, fortunately a much better world economically, where instead of just one strong economic nation, the nations of Europe, Japan and Asia, Canada and North America, all of these nations are strong economically, strong competitors, and as a result, it was necessary in these meetings for a negotiation to take place between equally strong nations insofar as their currencies were concerned." To be sure, Nixon was exaggerating the case, because no rivals to the U.S. dollar existed, and besides, the United States still had unequaled political and economic leverage. Nevertheless, he was signaling the U.S. recognition that the world had changed radically in the last twenty-five years and that with regard to the burdens it shouldered and the way it negotiated, the United States had to adjust. The devaluation of the dollar embodied all that.

AT A NEWS conference later that evening, Connally talked about how difficult the negotiations had been and how far the United States had come from the time it swore off any devaluation. "But we are pleased it's settled, everyone's pleased it's settled," he said. "We will return to a degree of stability."

The agreement was greeted in the United States with considerable elation. A significant currency alignment had been achieved. American exports would benefit. Foreign investment into the United States would expand. In both cases, the trade balance would be improved and jobs would be created. The strain on the U.S. gold reserve was over. The ground was laid for further negotiations on fundamental international monetary reform. The same could be said with regard to trade. Pressure had been exerted on the allies to expand their sharing of the burdens of defense spending, and NATO was poised to follow up.

Business sentiment was high; a top official at Morgan Guaranty Trust said, "After so many months and so many uncertainties, [the agreement] should bring a sigh of relief to bankers and businessmen everywhere." The chairman of IBM, T. Vincent Learson, praised the value of more predictable exchange rates for those doing international business. Robert Roosa, former treasury undersecretary under JFK and LBJ, praised the agreement as a "remarkably good job."

The agreement buoyed business spirits in Western Europe, too. "Essentially, the agreement removes a gnawing uncertainty that was already having the effect in European boardrooms of checking capital investment growth," wrote Clyde Farnsworth of the *New York Times*. There was a feeling that billions of dollars held by Western Europeans would flow back to the United States and into a booming stock market and other investment opportunities, he wrote.

Then, too, the world averted a major increase in trade protectionism. And Nixon's summitry with China and the USSR was put on a much stronger footing.

But the Smithsonian Agreement proved not to be all it appeared. When Volcker heard President Nixon's ebullient remarks on Saturday, December 18, at the National Air and Space Museum, he thought of what had just been negotiated and confided in a colleague, "I hope it lasts three months." In the event, his pessimism was justified, for the agreement would soon fall apart.

The Long View

Immediately after the Smithsonian Agreement was concluded on Saturday, December 18, the IMF gave it legal force among virtually all the countries of the free world. For Americans concerned with foreign policy, a sigh of relief could be heard at home and abroad that the hardball economic negotiations had not ruptured political alliances. Henry Kissinger wrote in his memoirs, "Nixon's unilateral decisions of August 15 had their desired effect. Allied cohesion had been strained but not broken." U.S. stocks soared, with money pouring in from Western Europe and Japan. Writing in the *Washington Post*, Hobart Rowen said, "The remarkable events set in motion on August 15 . . . could mark an upbeat chapter in U.S. economic history."

The agreement had its detractors, too. The *Wall Street Journal* cast doubt on the return to fixed exchange rates altogether, saying that for a rapidly growing global economy in which nations were ferociously jockeying for competitive advantage, what was needed was an agreement that allowed rates to fluctuate against one another. Within the administration, there was also frustration. Shultz bemoaned the lost opportunity to move to floating exchange rates. Connally's complaint was different. He believed that the size of the U.S. devaluation was inadequate. "We should have had a 36 percent revaluation from the Japanese, more than we got from the Germans, more from the French. And it all came back to haunt us," he said, looking back on the agreement. (One of the people I interviewed for this book, a keen observer of much that happened at Camp David and afterward, told me that Connally was bitter about having been boxed in by the foreign policy constraints that Nixon and Kissinger had put on him. According to this account, the treasury secretary wanted to keep the import tariff on longer and push the allies much harder, even if that meant the negotiations lasted for many more months. I can't be sure this is a totally accurate recollection, although it certainly seems plausible.) Paul Volcker held a similar view about the size of the currency realignment. "[The devaluation] was well short of what we felt was needed to restore a solid equilibrium in our external payments, even if we had succeeded in opening Japanese and European markets in trade talks," he later wrote.

Volcker was also dismayed about the fundamental intentions of the administration, the Federal Reserve, and some foreign governments. From the moment the agreement was concluded, he said, "What was lacking was the sense of commitment to make it all work." He meant that governments were not rigorously committed to buying and selling one another's currencies to maintain the new realignment; nor were they going to move their interest rates up and down to control capital flows. Volcker anticipated a dissipation of Washington's intense focus on currency issues. He wrote, "Nixon, like most presidents, resented the idea that his freedom of action might be limited by monetary difficulties. The only objective I heard him state about a reformed monetary system, and I heard him say it more than once, was that he just didn't want 'any more crises.'" Translation: if the Smithsonian Agreement quelled crises for a while, Nixon would move on to other things, and ensuring that the agreement worked wasn't one of them.

French president Pompidou also had apprehensions, especially about Washington. On Wednesday, December 22, just four days after the Smithsonian Agreement, he publicly aired his concerns in a television interview. Referring to commitments made at the Azores summit, Pompidou said that Nixon had promised to support the new exchange rates after realignment. To Pompidou, this meant the Treasury would buy greenbacks when their value sagged in international markets to ensure agreed-upon levels of value for the dollar. It also meant that the Fed would not allow interest rates to sink too low relative to the rates of other major countries, thereby diminishing the attractiveness of holding dollars. Pompidou would also have been happy if the United States expanded its controls on capital outflows both to prevent a flood of new dollars abroad and, by thus limiting the supply, to hold up the value of the greenback. But none of this happened. The French president complained, "The U.S. must first of all keep the moral undertakings it has made, that is to say, to restore its balance of payments by its own efforts and not merely by relying on the mechanical device of devaluation. There is no use in talking about a new international monetary system until the American balance of payments is back in equilibrium."

Many observers also stressed the critical importance of the longer-term negotiations for fundamental reform of the trading and finance that the G-10 had committed to negotiate. The *Washington Post* editorialized, "The new exchange rates for the dollar constitute a successful interim solution to America's troubles in the international economy. But the point to grasp is that they remain, most emphatically, an interim solution. The precise scale of the success will be determined by further agreements to follow." And the sheer number of questions that such long-term agreements would be forced to deal with showed how many difficulties had still to be addressed. "The underlying problems that created the monetary crisis of the last few years have not yet been decisively solved," wrote Leonard Silk of the *New York Times*. He listed a number of the unanswered questions, including: The United States could no longer back dollars with gold, so what would take the place of gold for foreign governments that held some $60 billion as reserves? How to ensure that exchange rates were defended either by the buying and selling of currencies or by changes in interest rate policies? How to make governments with balance-of-payments surpluses take actions to reduce those surpluses, just as governments in deficit needed to act to narrow their balance-of-payments gaps? How to make sure sufficient reserves existed in the global monetary system to accommodate an ever-greater amount of trade? When would negotiations over fundamental reform in the trading regime begin, and would they deal with the big controversies, such as agricultural subsidies in the EC or deeply protectionist regulations in Japan? Paul McCracken, who was about to retire, said that the world now faced a system problem, not just a dollar problem.

ON SUNDAY, NOVEMBER 14, 1971, a little over a month before the Smithsonian Agreement, Nixon's three-month wage and price *freeze* turned into a full-fledged program of wage and price *controls*. While these restraints held back escalating inflation during 1972—the critical national objective for the administration in an election year—there could be no avoiding the inherent difficulties with the more expansive and compulsory program. As the scope of the controls became broader and deeper, a sprawling bureaucracy emerged—exactly what Nixon feared and had promised would not happen. Fair treatment between business and labor, together with equitable handling of hundreds of specific requests for exceptions to the rules, complicated effective enforcement. The truth was, it was beyond even the most astute politicians and civil servants to manage the gigantic, complex U.S. economy with such controls. The program bred the illusion that it was possible to offset expansionary fiscal and monetary policies by smothering wages and prices with government decrees. The controls program went through several phases and even another freeze before being ended by Congress in 1974. Not surprisingly, when the controls were lifted, repressed prices were unleashed and inflation returned, eventually with a vengeance.

No one should have doubted that the president would pull out all stops to ensure his reelection in November 1972, just eleven months after the Smithsonian meeting. And what mattered most to his reelection prospects was national economic growth and especially lower rates of unemployment. The decisions of August 15, 1971, had boosted Nixon's image as a leader who took charge of the big domestic and international economic challenges. With that behind him, the president could neglect the broader picture of international monetary reform because the prospect of a dollar crisis had at least been postponed. Now, in direct contrast to his budget-restraining commitments announced in his August 15, 1971, speech, Nixon ran ever-bigger budget deficits. In fact, all the tax cuts that he had announced were implemented, but virtually none of the fiscal restraints. At the same time, the president relentlessly pressured Arthur Burns to lower interest rates.

As 1972 began, an economic resurgence had already taken hold in the United States. Housing and consumer spending were booming. Fiscal and monetary stimuli were adding gasoline to a fire. *BusinessWeek* captured Nixon's mind-set. "President Nixon has determined to lash the economic system to a gallop," it said. "His whips are the traditional ones—fiscal policy and monetary policy—but the way he is using it [*sic*] has no parallel in modern history." The magazine pointed out that the budget would have staggering deficits and that the Federal Reserve was trying to push money into the economy "by every means but assault and battery."

Nixon's pressure on Burns to lower rates has become a well-known story of how a president can undercut the independence of the Federal Reserve—so much so that no president after Nixon attempted to compromise the Fed's independence anywhere near as much (until, that is, President Trump, with his constant personal and public attacks on his Fed chairman, Jerome Powell). The Fed's loose monetary policy in the ensuing years is often blamed for much of the virulent inflation that characterized the mid- to late '70s and early '80s. In the year following the Smithsonian Agreement, low rates in the United States also contributed to more capital outflows to countries with higher rates, driving the dollar below the new level set by the devaluation. Whatever the motivations of Burns and the Fed governors, it was understandable why foreign governments believed that Washington was neglecting its commitments made in the Azores and at the Smithsonian. They also believed that the United States was intentionally driving the dollar lower in order to achieve a bigger decline in its value than it had been able to negotiate at the Smithsonian.

The United States' super-strong economic growth in 1972, Nixon's highly successful and widely publicized trips to Peking and Moscow that year, and substantial troop withdrawals from Vietnam all helped propel the president to a landslide reelection in November 1972, in which he carried every state except Massachusetts. From then on, as we now know, the administration descended into the tragic maw of Watergate and, slowly, then suddenly, ended its tenure with Nixon's resignation on Thursday, August 8, 1974.

The Western Europeans, in particular, were convinced that the United States lacked any urgency about monetary reform, allowing budget deficits to widen and interest rates to remain low. Barely two months after the Smithsonian, President Pompidou took Nixon to task for a second time. In a personal letter to the U.S. president about his disappointment with Washington, he wrote, "I feel obligated to describe to you . . . my uneasiness with regard to the evolution of the international monetary situation." Pompidou accused Washington of failing to defend the dollar and of moving sub rosa into a floating-exchange-rate regime, which was anathema to France and never part of the agreements in the Azores or at the Smithsonian. "It would be disastrous for the international monetary system and thus for the entire free world should the accords of December, whose historic character you yourself noted, became only a precarious pause along the path towards a new crisis." Nixon was unapologetic in his reply of Wednesday, February 16. He touted U.S. policies that were leading to faster economic growth and his wage and price controls, which were designed to squash inflation. In addition, he told Pompidou that many of the French leader's concerns would be taken up in a later stage of monetary negotiations. He then complained to Pompidou about Western Europe's foot-dragging when it came to dismantling trade barriers.

ON TUESDAY, MAY 16, 1972, John Connally resigned from office, believing that he had made as big an impact as he could as treasury secretary and hoping to further his political prospects, perhaps even to replace Spiro Agnew as Nixon's vice-presidential running mate later that year. He was replaced by the much-quieter and low-key George Shultz, who was formally put in charge of all departments and agencies having anything to do with economic policy. But the air was out of the balloon on global monetary reform. With the absence of Connally's drive and charisma; with Nixon's preoccupation with high-profile trips to Peking and Moscow; with the administration's intensifying focus on stimulating the domestic economy, with dollars flowing out of the United States in 1972 and 1973; and with Nixon's obsession with the election—with all that, it was no surprise that U.S. leadership in international monetary negotiations was sagging. Thus, the Smithsonian Agreement began to unravel soon after it was concluded.

On June 23, 1972, barely six months after the Smithsonian Agreement, the British made the first formal break with the parities that had been negotiated and devalued the pound yet again. At the rates they had negotiated in late December, the currency was overvalued, and the United Kingdom simply couldn't avoid ever-increasing trade deficits. By now, Nixon had lost interest in currency issues. On that day, the White House tape recorder captured the following dialogue.

Haldeman: Did you get the report that the British floated the pound last night?

Nixon: No. I don't think so, they have?

Haldeman: They did.

Nixon: That's devaluation?

Haldeman: Yeah. [White House aide Peter] Flanigan has a report on it here.

Nixon: I don't care about it.

Haldeman: You want a run down?

Nixon: No, I don't care. Nothing we can do about it.

Haldeman: [Federal Reserve chairman] Burns is concerned about speculation about the lira.

Nixon: Well, I don't give a shit about the lira.

That summer, newly installed treasury secretary George Shultz tried to resurrect U.S. leadership. He still wanted a system of floating exchange rates, but being a good soldier in the administration and understanding that the allies were not on the same free-market wavelength, he moved cautiously to promote a grand bargain in which currencies had substantially more flexibility to move up and down around a fixed relationship with one another. Shultz also introduced another idea that supported his laissez-faire approach to exchange rates: He proposed that those countries that wanted to float their currencies could do so, although the United States would not yet follow that route. Shultz's proposal constituted a serious set of arrangements that was considerably more flexible than the one agreed to at the Smithsonian but still a long way from the full-scale floating-rate regime he preferred. The time wasn't quite right for the United States, the center of the system, to abandon its fixed rate to gold, but in Shultz's mind, that day would come sooner rather than later. Still, his proposal made little progress. Virtually no government would gear its interest rate policy, or take any other measures, to defend the fixed parities agreed upon at the Smithsonian. Everyone's domestic priorities took precedence over what they were willing to do to preserve stable exchange rates.

Most important of all, the United States itself was doing nothing to uphold the system, disregarding all balance-of-payments considerations. As a result, all over the world the dollar was being sold and its value consequently declining, and less than four months after Shultz's proposals, Washington decided that it had to devalue the greenback once again.

On the evening of Wednesday, February 7, 1973, Paul Volcker left Washington on a secret whirlwind trip. In five days, he traveled 31,000 miles and consulted with officials in Tokyo, London, Paris, Rome, and Bonn. It was a far cry from the unilateralism of Sunday, August 15, 1971, for this time, Volcker talked confidentially and in advance with the United States' major economic and political partners about what Washington planned to do, and he solicited and received their acceptance before the United States took action. Upon his return on Monday, February 12, 1973, the United States announced a 10 percent dollar devaluation from its Smithsonian parity with gold. With this second American devaluation, no one could have much confidence in the durability of any agreement based on fixed parities. On Monday, February 12, 1973, the Japanese, the Italians, and the Swiss floated their currencies. By Friday, March 16, virtually all the Western European countries started floating again, and many foreign exchange markets were closed.

Then major unforeseen global disruptions came into play. From October 10–17, 1973, the Organization of the Petroleum Exporting Countries (OPEC) raised oil prices by more than 50 percent, and by the end of December 1973, the price had quadrupled. The centrality of energy to all economies made these price spikes the beginning of a series of economic dislocations:

inflation, recession, massive new borrowings and debts, changing competitive situations. Suddenly, major trade deficits began to appear all over the West and Japan and in oil-importing developing nations, and huge trade surpluses arose among oil-exporting countries ranging from Saudi Arabia to Venezuela. The oil exporters couldn't make use of all the income they were suddenly accruing, and at the same time, the oil consuming countries needed to finance their growing trade deficits. Thus, the OPEC surpluses would have to be recycled back to the deficit countries, a monumental challenge for international banks. The United States, the European Community, and Japan were in different positions regarding oil imports, and each reacted with different mixes of fiscal, monetary, and energy policies. Combined with the impact of the removal of wage and price controls in the United States and with Burns's refusal to raise interest rates, the precipitous and dramatic rise in oil prices led to a wave of inflation in the United States and abroad. This was aggravated by a worldwide crop shortage and soaring prices for many agricultural products. The combination of all these challenges placed a premium on domestic policies over any thought of returning to fixed parities, with the domestic discipline over fiscal and monetary policies that fixed exchange rates required. The global financial system had been shaken to its core.

APPROXIMATELY TWENTY-FOUR MONTHS months after it was concluded, the Smithsonian Agreement, characterized as it was by fixed exchange rates and the implicit understanding that each nation would adjust its home policies to support those rates, was a dead letter. Floating exchange rates became the norm. Paul Volcker later reflected, "In the process, for better or worse, the sense of fixed structure, stability and order that for a while had characterized the world economy in the context of the Bretton Woods system seemed to disappear."

On Thursday, May 9, 1974, Treasury Secretary Shultz resigned, in large part because Nixon insisted on continuing wage and price controls over his objections. He was replaced by William Simon, a former Wall Street bond trader and rigid free-market advocate. Unlike Shultz, Simon had little patience for finding a common interest among nations. At the annual IMF meeting in September 1975, he minced no words about the U.S. position. "The right to float must be clear and unencumbered," he said, reinforcing what everyone had grown to understand: the United States was not going to support a return to fixed rates, even with a more flexible band around them.

Four months later, from January 7 to 8, 1976, the "Jamaica Accords"—they were concluded in Kingston—became the successor to the Smithsonian Agreement and legitimized floating exchange arrangements under the IMF Articles of Agreement. The formal price of gold for monetary purposes—which stood at $42.22 an ounce after the second U.S. devaluation in 1973—was abolished, and henceforth the value of gold would be whatever the market determined. This meant that gold, for all practical purposes, no longer played any special role in monetary affairs—no more special than any other asset that a central bank might want to hold in reserve. In addition, new rules for floating or for a better balance of obligations between surplus and deficit countries were devised but were not accompanied by any strong enforcement authority.

The oil crisis, worldwide inflation, and recession had overwhelmed efforts to reform the monetary system. As economist John Williamson wrote a year later, Jamaica represented "a decision to learn to live with a non-system that [had] evolved out of a mixture of custom and crisis over the preceding years." Governments could do pretty much as they pleased regarding their exchange rates, the volume and compositions of their reserves, and the way they went about adjusting their deficits or surpluses. After more than four years, the search for a set of rules for a new monetary order—forced by Nixon's August 15, 1971, announcements—had failed.

Some journalists put it more diplomatically: "It will be, in other words, a very permissive system, in which standards of behavior and styles of financial management are likely to vary considerably from country to country, with no supranational power to ban either floating or pegging," wrote Leonard Silk in the *New York Times*. Treasury Secretary William Simon put it this way at a cocktail party following the end of the conference, perhaps reflecting the fact that the G-10 had gone as far as it could or wanted to: "All is well that ends."

CAMP DAVID LED to the Smithsonian, which led to the Jamaica Accords and the legitimization of a new agreement to allow currencies to float with very few rules. It seemed to be the only feasible framework for the extremely volatile times, and in any event, it was the only system that nations at the time would accept. But floating did not prevent many more dollar crises in the ensuing years. In 1978, during the Jimmy Carter presidency, the low level of confidence in the administration's policy at home and abroad drove the dollar down so low that a massive international rescue operation was necessary to raise its value. In 1985, during the second term of Ronald Reagan, the dollar became so strong, in large part because of sky-high interest rates imposed to kill inflation, that a concerted effort on the part of the United

States and other major industrialized nations was mounted to push down its value. The negotiations became known as the Plaza Accord, as they were concluded at Manhattan's Plaza Hotel.

As of 2020, the Plaza Accord was the last major successful multilateral effort to coordinate currency values. That agreement worked for a while, and it averted a major bout of protectionism, not to mention a potential political rupture between the United States and its allies. Nevertheless, markets ultimately took over, and the impact of the Plaza Accord was dissipated. More crises ensued.

In the 1980s, Latin American governments that had borrowed heavily in dollars went bankrupt when the value of the dollar soared and they couldn't repay their debts. In 1987, the United States experienced a stock market meltdown, in part because of disputes over trade and currency issues between the United States and West Germany. In 1995, to prevent a broader banking catastrophe, Washington and the IMF had to bail out Mexico, which couldn't pay its dollar-denominated debts. In 1997, currency disturbances that began in Thailand spread like wildfire throughout Asia and into emerging market economies on other continents, sending the dollar on a roller-coaster ride. This was followed by a Russian default on dollars owed to U.S. banks that, until quelled, threatened to do major damage to Wall Street.

The biggest global financial crisis since the Great Depression arrived in 2008. The effects lingered for at least a decade and led to another long crisis that enveloped the eurozone, centering first on Spain and Portugal and then gravitating to Greece.

The long-term impact of the coronavirus pandemic could constitute another global economic and financial catastrophe, as the global economy slows to a crawl, governments take on unprecedented amounts of debt, central banks turn themselves into currency-printing presses, bankruptcies multiply, and the world's productive capacities and workforces take a generation to recover. On October 15, 2020, Kristalina Georgieva, the managing director of the IMF, said as much in addressing the eventual destructive impact of COVID-19. "Today we face a new Bretton Woods moment," she told a meeting of the world's top financial officials.

One thing is for sure: There will be more crises. Some may come from failures within the financial system, such as reckless risk taking combined with inadequate cushions of bank reserves. Some may come from challenges to dollar supremacy from China or a strengthened eurozone. Some may originate from digital currencies or massive cyberattacks on the financial system itself. And some may result from extreme levels of indebtedness in the post-COVID era combined with rising interest rates and debt-servicing costs. But no matter the reason, the U.S. dollar will be at the center of any global disturbance, and managing the greenback will be part of the challenge and the solution.

The Weekend in Retrospect

The weekend during which a small group of men decided to sever the greenback from its gold moorings was a critical turning point for the modern global economy. The Nixon administration's actions over those three days led to the most significant structural changes regarding the dollar and the international monetary system between the 1944 Bretton Woods Agreement and today. This was no small thing. After all, the dollar was at the heart of the astounding recovery of Europe and Japan after World War II. It was a major ingredient in the unprecedented prosperity the United States enjoyed for three decades after the war. It was a potent symbol of American industrial and military power across the globe. It was the standard against which all other currencies were valued. The greenback was at the crossroads of foreign and domestic policy, affecting the United States' relations with the world; driving a good deal of international commerce; and influencing the country's economic growth, employment, and industrial structure. Moreover, the decisions at Camp David were not only of historic proportions but represented the best that imperfect men could achieve operating in a cauldron of political and economic change at home and abroad.

Of course, no set of such sweeping policies are without legitimate controversies. So, before explaining why the August 15 proposal should be seen as a highly positive turning point for the United States and the world, let's evaluate some legitimate criticisms.

WAGE AND PRICE controls were a fiasco. Embedded in the weekend decisions was a major mistake: the imposition of wage and price controls. To be sure, these constraints were wildly popular at the time, and they did work to suppress price increases for a year or two. Burns supported them because he felt that conventional monetary policies wouldn't work. Connally was in favor because he saw the short-term political value of their impact. Volcker held his nose and agreed to them because he was convinced that the desirable alternative—that the Fed would raise interest rates to hold down inflation—would not happen. Shultz, an ardent free-market advocate, was adamantly opposed to them for ideological and practical reasons, although he ended up acquiescing and even administering them, fooling himself that they would be extinguished fairly quickly. He later described the controls in this way: "Bad medicine at a bad time."

Shultz would be proven right. The controls were too hard to administer fairly and led to an unwieldy bureaucracy. They gave the administration the illusion that the government could then pursue easy money and lax fiscal policy—which it did. The biggest problem was that once the controls were removed, pent-up pressures to raise prices and wages were unleashed, resulting in the very inflation that the Nixon administration had been trying to avoid and fueling escalating inflation that lasted many years. As noted, these pressures were not offset by the Fed's willingness to raise interest rates. Richard Nixon, acknowledging the policy's failure, delivered an appropriate epitaph on the wage-price fiasco in his memoirs. "The August 15, 1971, decision to impose [wage and price controls] was politically necessary and immensely popular in the short run," he wrote. "But in the long run I believe it was wrong. The piper must always be paid, and there was an unquestionably high price for tampering with the orthodox economic mechanisms."

Vital aspects of American competitiveness were neglected. If wage and price controls were a sin of commission, failure to focus on the deepening economic and social causes of the decline in U.S. competitiveness was a sin of omission. In wanting to devalue the dollar and in aggressively pursuing more trade openings abroad, Nixon aimed to improve U.S. competitiveness. But Washington failed to deal with many of the deep-seated causes of America's problems, those that went beyond the dollar and trade. The administration made a big mistake in equating competitiveness only with the balance of payments (especially bilateral trade deficits) and not with deeper economic and social causes.

A major part of a broader competitiveness agenda had been advocated by Peter Peterson. In his 1971 study, "The United States in the Changing World Economy," he focused heavily on investing for the future, particularly in technology and R&D and in the retraining of workers who were hurt by imports. But Peterson's recommendations were essentially ignored. And, if anything, an even wider agenda than what Peterson advocated would have been in order. That would have entailed retooling skills for workers impacted by the offshoring of American production, and also for those whose jobs were eliminated by rapidly changing technology. The right policy prescription would have included continual upgrading and modernization of the United States' network of roads, seaports, and airports. Other critical reforms would have embraced investments in secondary education and in a strong and resilient social safety net to help the overall population better weather the ups and downs of the global market.

America acted too unilaterally. A third criticism of the strategy pursued by the Nixon administration is that it opted for restructuring the global framework of fixed currencies with no advanced consultation with close allies. It was, in other words, an act of raw unilateral power at a time when the global economy in which so many countries were now involved required more multinational cooperation. After all, West Germany, Great Britain, France, Italy, Japan, and the other U.S. trading and defense partners all had fundamental interests at stake. To rub salt into the wound, the argument goes, Nixon imposed an across-the-board tariff that was a breach of the free-trade policy the United States had been zealously espousing around the world. The impact of this unilateralism was to create disarray in markets and in critical political alliances. Such turmoil made little difference to Connally, who even relished the conflicts as a sign the United States was standing up for its interests. Arthur Burns opposed such arrogance, and Paul Volcker was worried about it. Kissinger worked hard to overcome the tensions it created. As for Nixon, he was on all sides of the issue at one time or another.

In reality, however, there was no effective way to start a multilateral negotiation without Washington's administering a shock in advance. The United States needed to demonstrate that there would be no choice but to negotiate, and to do so quickly. Without such a demand, negotiations would likely have been resisted by the allies. After all, they were prospering under the existing system. They were being asked to revalue their currencies, an act which would disadvantage the exports on which they were so highly dependent. And the United States was not giving them much in return, except for some pledges to manage its economy better, which the Europeans and Japanese thought it should be doing anyway. Also, it would have been impossible to negotiate a currency realignment without markets getting wind of the discussions and inevitably causing chaos in global trade and investment.

The fact is, the international monetary system is always evolving. The geopolitical context changes. The markets themselves grow bigger and more complex. A major challenge to the world is how to adapt to these changes. Can they be done smoothly, or are major disruption and political conflict inevitable? In the summer of 1971, Washington decided that brute unilateral force was the only way to effect change. It was a gamble, to be sure, but it worked.

The administration lacked a long-term strategy. Another possible shortcoming of Washington's approach was that it lacked a genuine strategy beyond the announcement of August 15, 1971. That is certainly what Peter Peterson kept saying. Among the reasons for this omission was that the Nixon administration was not of one mind about what the post–August 15 currency arrangements should be. The president's advisors knew that major change was needed to take the burden off the dollar, but there was no consensus on the precise nature of how to do that over the longer term. Volcker and Burns were in favor of a currency realignment and then a return to fixed rates. Shultz and McCracken favored a permanent floating of currencies. Connally was riveted on domestic politics, which favored keeping all options open. Kissinger was preoccupied with the cohesion of alliances; he wanted to settle sooner rather than later, regardless of the exchange rate regime. Nixon, who, like Kissinger, just wanted any agreement, seemed to back everyone at one point or another. Granted, it was a messy process, but when you examine them carefully, most big government decisions are.

Besides, Washington couldn't have had a clear view of how other countries would react. It needed some flexibility to adjust and respond. It did the best it could under these circumstances.

The Camp David weekend did not achieve real change until the Jamaica Accords of 1976. It was too long and tortuous an ordeal. There can be no denying this. But the decisions made between August 13 and 15, 1971, broke the dam of massive political inertia among the United States and its allies. And while the 1976 Jamaica Accords gave floating rates official status, in fact most countries started floating within fifteen months after the Camp David meeting. Would it have been better to move from fixed to floating exchange rates cleanly and more quickly? It's not clear. Perhaps a series of half steps, plus some experimentation, was politically necessary. Maybe the transition was essential given the large number of significant countries involved, each with its own political system and with different economic and social pressures.

U.S. policies unleashed a decade of inflation. Some critics of the weekend allege that in addition to the ill-advised wage and price controls, the freeing-up of exchange rates was responsible for the period of high inflation in the United States and Western Europe for so much of the 1970s and early '80s. Ten years ago, on the fortieth anniversary of the Camp David weekend, for example, financial commentator Roger Lowenstein wrote, "The Nixon Shock was a central cause of the Great Inflation." Shortly afterward, economic historian Lewis Lehrman, writing in the *Wall Street Journal*, pointed to how much the dollar had depreciated against gold since the Camp David weekend and attributed "the worst inflation in American history" to the Nixon decisions.

True, inflation did soar in the 1970s, causing widespread damage to the economy and the lives of millions. If the dissolution of the dollar–gold pegging was to blame, surely it was only one factor, and probably a minor one at that. Price pressures from escalating oil prices were a main culprit. So was the lack of reinforcing monetary policy—resulting from Nixon's sustained pressure on Arthur Burns to keep interest rates too low throughout 1972 to help ensure his reelection. If it had had the political courage to accept the cost of a severe slowdown in economic growth, the Nixon administration could have slayed inflation by jacking up interest rates. That, in fact, is what Volcker eventually did when he became Fed chairman in the Carter administration. His actions led to a deep recession, but inflation came down dramatically. Floating exchange rates didn't stop him. And when high interest rates in the United States caused the dollar to soar, the Reagan administration brought the G-10 countries together and negotiated a temporary currency realignment.

The administration removed the one and only anchor for the international monetary system. Some critics of the Camp David weekend bemoan the removal of gold as the anchor for the dollar and, by connection, for all currencies. From ancient days, gold or some other metal had been used either as money or as tangible collateral for it. Those who advocate that currencies be backed by hard assets worry that without such linkages, governments would act in an undisciplined way, print money with little restraint, and cause the debasement of the currency via rampant inflation. Although inflation did soar in the 1970s, it has not been a problem, to say the least, for the past three decades.

The same people who bemoan the delinking of the dollar from gold often point to the 1930s, a period when the gold standard broke down and unhinged currencies led to destructive protectionism. And it is true that after August 15, 1971, the world entered an era of fiat currencies—currencies with no backing by anything physical. This didn't make much of a difference in the world economy, however. We had periods of slump and rapid economic growth before and after the summer of 1971. Financial crises existed before and after. Following Camp David, the world economy continued to grow, and trade flourished. In any event, the debate over the gold standard was becoming moot by the summer of 1971. There just wasn't enough of the precious metal in central banks to support the rapid increase in global commercial transactions, and no one could figure out what kind of collateral would take gold's place. The great risk of backing the currency with gold or another metal is that there won't be enough supply of that commodity to support the expansion of commerce. That situation could constrain the growth of the world economy and lead to deflation. When you have deflation, prices go down. Investors stop investing because they worry about diminishing returns. Consumers stop buying because they think prices will be cheaper tomorrow. The entire economic machine stops working. In any event, today, when we might be on the verge of a world of digitalized currencies, fixing the dollar to gold or any metal seems not just outdated but quaint.

It was all about the election. Finally, it can be argued that Nixon and Connally were driven to reform international monetary relations solely by the imperatives of the upcoming presidential election. No counterargument here, but it would be naïve not to admit that every administration moves to the rhythm of electoral politics. Besides, national security and economic burdens on the United States were also pushing Nixon to act as he did. Looking back, even without the election, one sees he didn't have much choice but to push hard for a formula that would take pressure off the U.S. currency. The gold just wasn't there.

WHEN THE ENTIRE record is reviewed, however, the weekend at Camp David can be seen as an impressive achievement. There are at least four reasons:

Gold had become an untenable burden for the United States and floating exchange rates were good for America. It followed that the eventual outcome of floating exchange rates served the United States and the world much better than fixed rates. True, Washington originally set out to maintain fixed exchange rates, albeit with new, realigned values and with more flexibility for currencies to move up and down against one another. It is the case, also, that this plan was never achieved and that the end result was today's world, in which major currencies float against one another with minimal intervention by their governments. This was a defeat for Volcker and Burns, and a victory for Shultz and McCracken. In the complicated world of international economics, it would have been impossible to be able to bridge two eras with one bold stroke. Too many countries were involved, not to mention highly sophisticated markets. Washington moved by trial and error, as it had to, and it finally arrived in the right place.

Here's why an exchange rate regime that allows nations to float their currencies against one another served U.S. interests: Fixed rates assumed that governments would gear their fiscal and monetary policies to ensure exchange rate stability. For example, suppose the dollar was sinking in world markets and that greenbacks were moving out of the United States because the American economy was slumping. In that situation, the Federal Reserve would be obligated to raise interest rates to attract dollars back to the United States. But higher rates would weaken the economy further. This would mean less priority to economic growth and job creation at home. In essence, fixed rates implied more attention to the expansion of international trade over domestic concerns. For a nation the size of the United States, in which trade constitutes a relatively small percentage of its GDP, this requirement is politically impossible because America needs the flexibility to give priority to internal economic and social needs. (In the early 1970s, total trade as a percentage of GDP was about 5 percent. Today it is closer to 20 percent. But in each case, America's dependence on trade was and is much less than countries such as Canada, the United Kingdom, Germany, and Japan.)

To be sure, Washington never did play by the rules and use monetary and fiscal policy to manage the dollar's exchange rate. This refusal was a huge problem for other nations that looked to the United States to take more responsibility for its policies.

In any event, the international acceptance of floating rates, which by definition did not require a country to defend its currency with adjustments in domestic policy, gave the United States the official permission it needed to pursue the policies it wanted at home. It would be a tragedy, of course, if the lack of international rules caused the United States to abandon all discipline and run up debilitating trade and budget deficits as if they didn't matter. That would not be the fault of the Camp David decisions, however, but of how Washington, both the executive branch and Congress, implemented them.

In addition, a world of fixed exchange rates leads to too much pressure on currency relationships at a time when the global economy is in constant change. Whether the shocks are in the form of financial crises, booms and busts, trade wars, man-made or natural disasters, or political tensions, countries need more flexibility to manage their currencies without the political trauma and economic impact of frequent devaluations or revaluations. Floating exchange rates allow countries to accommodate big changes in the world economy in a smooth and undramatic way.

The U.S. dollar continued to reign supreme, with all the attendant economic and foreign policy benefits. Despite the choppy waters for the dollar after August 15, 1971, the dollar still accounts for about 60 percent of all foreign exchange reserves in central banks and is far the most important currency in international trade and lending. U.S. Treasury bonds and bills never stopped being considered the world's safest financial asset. As a result, the dollar has been in demand everywhere, allowing the United States to finance its deficits with considerable ease and at low interest rates. In foreign policy, the dollar has remained a central part of the United States' image and soft power.

It may well be the case that the dollar will face new challenges from the Chinese yuan, the euro, or even a new world of digital currencies. But floating rates will not be the reason. The pre-eminence of the greenback depends on many other things: on the strength of the U.S. economy, on the breadth and depth of U.S. capital markets, on the confidence that investors and traders have in the integrity of critical institutions such as the Federal Reserve System, and on the rule of law in the country. The role of the dollar ultimately depends, too, on the quality of the government's policy—on the skills, experience, and knowledge of its leadership. Will the Treasury, the Federal Reserve, and the financial regulators manage the economy not just for the short term but with concerns for the more distant future? Will they take account of America's interdependence with the rest of the world? The course of the dollar will rest on the answers to questions such as these, much more than on the nature of the exchange rate regime.

Dangerous protectionist pressures were quelled. The decisions of that 1971 weekend led to developments in trade that diverted the world from a brewing disaster in the form of a protectionist spiral reminiscent of the 1930s. The Camp David decisions resulted in more balanced international trade. More realistic exchange rates were the first step. Following that, Washington led a new round of global trade negotiations—the so-called "Tokyo Round"—that addressed non-tariff barriers in a major way for the first time in history, negotiations that were begun in 1973 and concluded in 1979. At the same time, it embedded in its own legislation a number of defensive measures to combat illegal foreign trade practices and destructive surges in imports. This caused Congress to feel the United States was better equipped to resume its traditional open-trade policy.

The spirit of Bretton Woods was preserved. Amazingly, the Camp David decisions did not undermine cooperation among the allies; in fact, after an initial tense period, they strengthened it. The August 15 shock became a wake-up call that big changes in the world economy were needed and that they would require everyone's participation. Thus, the weekend led to years of ever-closer collaboration among finance ministers, central bankers, trade ministers, chiefs of development agencies, and heads of state. International institutions such as the World Bank and the IMF became more central to the global economy.

In April 1971, Peter Peterson described what the United States should be aiming for. "The new system must be characterized by shared leadership, shared responsibility and shared burdens," he said. "It will be a system which fully recognizes, and is solidly rooted in, the growing reality of a genuinely interdependent and increasingly competitive world economy whose goal is mutual, shared prosperity." The Camp David decisions were a prime catalyst for advancing this vision.

In fact, in the decade following the Camp David weekend, more in-depth thinking was done about the management imperatives for the rapidly changing world economy than at any time since. In think tanks and international fora around the world—from the Brookings Institution to the United Nations, from the Paris-based Organisation for Economic Co-operation and Development (OECD) to the Geneva-based General Agreements on Tariffs and Trade (GATT)—report after report generated ideas about the future shape of international monetary relations, including trade, economic development, the environment, and the management of international organizations.

It is doubtful that such a flowering of multilateral cooperation would have occurred if the rigidity of fixed exchange rates had not been broken, for currencies have been the prime enabler—the circulatory system, so to speak—of international economic activity.

In essence, the August 13–15 weekend used what we would today call "America-first" tactics not to bring down the existing structures of international cooperation, but to rebalance power within them. The purpose was not to destroy the alliances and organizations that were built after World War II, but to spread the responsibilities among key nations for making those alliances and organizations work. The goal was not to destroy but to adjust and modernize.

WAS THIS SUCCESS a result of historical trends or the talents of the Nixon administration? The answer is: both. Certainly, the political and economic pressures acting on Washington had been building for at least a decade, and they reached a fever pitch in 1971. At the same time, Nixon's economic team represented an important variety of experiences, skills, and perspectives that could push the United States in the right direction. Connally, in particular, forced clear decisions. Without him, the United States might not have delivered such a targeted and dramatic shock to the international system. Nevertheless, Nixon and Kissinger made sure that foreign policy considerations were never out of sight. Shultz and McCracken brought a free-market perspective; Peterson, a broad strategic view. Volcker brought detailed knowledge of how global finance actually functioned. Occasionally, they may have disagreed, gone behind one another's back, sought the president's favor to undercut one another. In the end, though, a sense of public service motivated all of them. It would be hard to envision a more capable group for the challenges they faced.

IN THE END, the Camp David weekend was part of a fundamental transition that the United States had to make from one era to another. It constituted a bridge from the first quarter century following the war—when the focus was on rebuilding national economies that had been destroyed and on reestablishing a functioning world economic system—to a new environment where power and responsibility among the allies had to be readjusted, with the burden on the United States being more equitably shared and with the need for multilateral cooperation to replace Washington's unilateral dictates.

We are today again in a major transition of such magnitude. It was not long ago that most of the world accepted increasing globalization and expanding international economic cooperation as the order of the day. Now, though, we are headed in a different direction, one that gives priority to rebuilding the engines of growth and the social safety net at home over the international agenda. We are living in a world in which the two biggest and most important nations, the United States and China, are headed for a Cold War or maybe something worse. Other elements of a major transition are evident—a gigantic economic restructuring requiring trillions of dollars of public investment as a result of the worst pandemic in a century; a massive change from an era of fossil fuels to a new age of a cleaner climate; a rapid demographic transition from a predominately white society to a multiracial country; and a technological transition in which artificial intelligence and machine learning will change virtually everything.

All these trends and imperatives will put severe stress on the operation of the global economy in general and the international currency and monetary system in particular.

There is another parallel between 1971 and today. At Camp David, the Nixon administration deliberately delivered a big shock to what was then called the free world. It believed this was the only way to get other countries to accept the need for fundamental changes in finance and trade. It is no exaggeration to say that the Trump administration delivered an even bigger shock in the form of subverting long-standing alliances, withdrawing from international agreements and organizations, and generally showing disdain for all forms of international cooperation. In my view, these were ill-advised and destructive measures. Nevertheless, the Trump shock creates an opportunity for the current administration to pick up the pieces and establish a new set of global arrangements to deal with the formidable agenda ahead. That includes rebuilding the American economy so that it benefits all its citizens, dealing with big powers such as China and Russia, and managing international problems ranging from the need to improve trade and financial arrangements to confronting climate change and strengthening the infrastructure for public health. President Biden faces no greater set of challenges.

AUTHOR'S NOTE

Over the last three decades, I have written a number of books on global themes: competition among America, Germany, and Japan; the rise of big emerging market nations such as China, India, and Brazil; the role of business leaders in society; and the story of globalization through the lives of ten people. Despite this focus on the big picture, I have secretly admired authors who could zero in on just one specific event, make it come alive, and use it as a springboard to give readers the larger historical significance. Writing this book was my first attempt to do just that.

I chose the weekend of August 13–15, 1971, because it overlapped with my professional interests since the early 1970s. I began my career in the Nixon administration in late 1973, two years after the Camp David meeting. Ever since, I have been immersed in international economic and financial policy—in the Nixon, Ford, Carter, and Clinton administrations, as an investment banker engaged for many years in international finance, and as a dean and professor at the Yale School of Management teaching courses on the global economy.

In fact, ever since I was in college I had a good idea of what I wanted my professional life to be about. I was interested in the intersection of public policy on the one hand, and finance and business on the other. I wanted to work in the space where international economics and foreign policy overlapped. And I wanted a foot in both the "real world" and the world of ideas. I have had the enormous fortune of doing all these things. Moreover, in this book, I had the pleasure of writing about other men who did them, as well.

Writing about economic issues for a general audience is a challenge in itself, but encasing them in their historical, political, and foreign policy contexts makes the task even more difficult. I tried to keep the economic dimensions as nontechnical as possible, while not sacrificing accuracy, and to portray them as well in the imperfect and oversimplified way they were seen by Nixon and his team. In this regard, one of my hurdles was to describe what each of the men at Camp David brought to the table by way of knowledge and experience, the factors they weighed and those they discarded either consciously or out of ignorance, and the lens through which they regarded the problems at hand.

I gained a lot of insight from the extensive interviews I conducted (see "Author Interviews," page 421). However, fifty years is a long time ago. In my session with him, George Shultz admonished me not to rely too much on what he said. You tend to remember the good things, while repressing the stuff you're not proud of, he told me. His warning made me careful to balance what my interviewees recalled with the written record.

As I wrote this book over the past few years, I felt some pride in the U.S. government and how it had managed the events I was describing. I was telling a story of highly dedicated government officials who acted in the public interest as best as anyone could have expected them to do. As someone who served in several administrations, and who always felt surrounded in government by many talented and dedicated people, I found the Camp David group, and many who worked with these men on the sidelines, especially inspiring in terms of the talent and experience they brought to the table and their genuine concern for the United States' problems and its future. As a result, in the four years I worked on this project, there was never a time when I didn't look forward to the research, the writing, the editing, and all else that is part of producing a book.

ACKNOWLEDGMENTS

I would like to thank the following people who provided extremely helpful comments on early drafts: Robert Zoellick, Fred Bergsten, Marc Levinson, Benn Steil, Edwin Truman, and Susan Schwab.

I am deeply indebted to the many men and women whom I interviewed. (See page 421.)

My editor at HarperCollins, Jonathan Jao, was invaluable from the first day we worked together on this book. In the most gracious way, he coached me to move from a nine-hundred-page first draft to the final book version, helping to shape the tone and structure and to eliminate material I hated to part with. An ideal editor. In addition, Sarah Haugen at HarperCollins was exceptionally helpful in overseeing all the steps involved in turning a manuscript into a book.

My agent, James Levine of the Levine | Greenberg | Rostan Literary Agency, was also superb. It is hard to imagine how anyone could be more helpful. Jim read every draft, provided thoughtful judgments and comments big and small, and was available around the clock for anything I wanted to talk about. An ideal agent.

I owe many thanks to Stephanie Posner, a graduate student at Yale, for extensive fact checking, proofreading, and editorial assistance. She has been a delight to work with.

And most of all, I am indebted to Kelly Jessup, my special assistant at Yale, who helped me with research, document retrieval, communication with presidential libraries, setups for interviews, editorial suggestions, and everything else that went into the book, including extensive work on the notes and bibliography.

Following are the names and positions of the principal figures in the story. The names of those who attended the Camp David meeting on August 13–15, 1971, are marked with an asterisk.

Barber, Anthony—Chancellor of the Exchequer, Great Britain, 1970–74

Bergsten, C. Fred—Deputy for International Economic Affairs, National Security Council Staff, January 1969–June 1971

*Bradfield, Michael—Deputy Counsel, U.S. Treasury, 1968–75

*Burns, Arthur F.—Chairman, Federal Reserve Board, January 31, 1970–March 8, 1978

*Connally Jr., John B.—Secretary of the Treasury, February 11, 1971–June 12, 1972

*Dam, Kenneth W.—Assistant Director, National Security and International Affairs, Office of Management and Budget, 1971–73

*Ehrlichman, John D.—Counsel to the President and Assistant to the President for Domestic Affairs, January 20, 1969–April 30, 1973

Giscard d'Estaing, Valéry—Minister of Economy and Finance, 1969–74; President of France, 1974–81

Gyohten, Toyoo—Japanese Finance Ministry, 1955–89, Vice Minister of Finance for International Affairs, 1986–89

*Haldeman, H. R.—Nixon's Chief of Staff, January 20, 1969–April 30, 1973

Hormats, Robert D.—Senior Staff Member, National Security Council, 1969–77

Javits, Jacob K.—U.S. Senator (R) from New York, January 3, 1957–January 3, 1981

Kissinger, Henry A.—National Security Advisor to the President, January 20, 1969–November 3, 1975

*McCracken, Paul W.—Chairman of the Council of Economic Advisers, February 4, 1969–December 31, 1971

*Nixon, Richard M.—37th President of the United States, January 20, 1969–August 9, 1974

*Peterson, Peter G.—Assistant to the President for International Economic Policy and Executive Director, White House Council on International Economic Policy, January 1971–February 1972

Petty, John R.—Assistant Secretary of the Treasury for International Affairs, May 1968–February 1972

Pompidou, Georges J. R.—President of France, June 20, 1969–April 2, 1974

Reuss, Henry S.—Member, U.S. House of Representatives (D) from Wisconsin, January 3, 1955–January 3, 1983, and Chairman of the Joint Economic Subcommittee on International Exchange and Payments

Ribicoff, Abraham A.—U.S. Senator (D) from Connecticut, January 3, 1963–January 3, 1981

Roosa, Robert V.—Undersecretary of the Treasury for Monetary Affairs, 1961–64; Partner, Brown Brothers Harriman, 1965–91

*Safire, William—speechwriter for President Nixon, 1969–73

Schiller, Karl A. F.—Minister of Economic Affairs of Germany, 1966–72, and Finance Minister, 1971–72 (overlapping positions)

*Shultz, George P.—Secretary of Labor, January 22, 1969–July 1, 1970; Director of the Office of Management and Budget, July 1, 1970–June 11, 1972; Secretary of the Treasury, June 12, 1972–May 8, 1974

*Stein, Herbert—Member and later Chairman, Council of Economic Advisers, January 1969–August 1974

*Volcker Jr., Paul A.—Undersecretary of the Treasury for Monetary Affairs, January 20, 1969–April 8, 1974

*Weber, Arnold R.—Associate Director, Office of Management and Budget, 1970–71; also Executive Director, Cost of Living Council, for ninety days in late 1971

*Weinberger, Caspar "Cap" W.—Deputy Director, Office of Management and Budget, July 1970–May 1972

NOTES

Introduction

1 At 2:29 p.m. on Friday, August 13, 1971: All specific dates and times come from the *Presidential Daily Diary, 1971*, White House Central Files, Staff Member and Office Files, Office of Presidential Papers and Archives, Richard Nixon Presidential Library and Museum, Yorba Linda, CA [hereafter "RMN Library"]. For this entry, see August 15, 1971, Box RC-8, folder "President Richard Nixon's Daily Diary August 1, 1971–August 15, 1971."

2 The space was cramped: All information on the helicopter comes from a personal tour of it at the Nixon Library and Museum on August 14, 2019, plus Book #31, White House photo collection in RMN Library.

2 "This is the biggest step in economic policy": Christen Thomas Ritter, "Closing the Gold Window: Gold, Dollars, and the Making of Nixonian Foreign Economic Policy" (diss., University of Pennsylvania, 2007), 271.

3 "This could be the most important weekend": William Safire, *Before the Fall: An Inside View of the Pre-Watergate White House* (Garden City, NY: Doubleday, 1975), 510.

4 other currencies, such as the British pound, the West German mark: The euro did not exist in those days, and each European country had its own currency.

6 The extreme trade protectionism that resulted: For a more complete picture of the interwar period, see C. Fred Bergsten, *The Dilemma of the Dollar: The Economics and Politics of United States International Monetary Policy* (New York: New York University Press, 1975), 46–79; Ragnar Nurkse, *International Currency Experience: Lessons of the*

Interwar Period (New York: League of Nations, 1944), 113–36; and Douglas A. Irwin, *Clashing Over Commerce: A History of U.S. Trade Policy* (Chicago: University of Chicago Press, 2017), 330–413.

8   "I want to make it . . . clear": John F. Kennedy, "A Special Message from President Kennedy to Congress," July 18, 1963, in Robert V. Roosa, *The Dollar and World Liquidity* (New York: Random House, 1967), 319.

8   "The dollar is, and will remain, as good as gold": Lyndon B. Johnson, in Roosa, *The Dollar and World Liquidity*, 348.

10 Thus, Washington had only 25 percent: The gold figures I'm using throughout come from United States Gold Commission, "Volume I," in *Report to the Congress of the Commission on the Role of Gold in the Domestic and International Monetary Systems*, March 1982, and Richard Nixon, *International Economic Report of the President 1974* (Washington, DC: U.S. Government Printing Office, 1974), unless otherwise indicated. They are consistent with multiple reports issued by the various administrations. To the extent there is any room for disputes or inconsistencies, it has to do with the magnitude of foreign liabilities in 1971. For that year I am using only liabilities held by foreign central banks and governments, which was the convention in the 1960s. If you include *all* dollars held abroad, then the shortfall between U.S. gold reserves and the total liabilities in 1971 is much greater.

15 references in the memoirs of some of the Camp David participants and in histories of international finance: For example, see Richard M. Nixon, *RN: The Memoirs of Richard Nixon* (New York: Grosset and Dunlap, 1978), 516–22; Safire, *Before the Fall*, 509–28; Henry Kissinger, *White House Years* (Boston: Little, Brown and Company, 1979), 949–67; George P. Shultz and Kenneth Dam, *Economic Policy Beyond the Headlines* (Stanford, CA: Stanford Alumni Association, 1977), 109–33; Paul Volcker and Toyoo Gyohten, *Changing Fortunes: The World's Money and the Threat to American Leadership* (New York: Times Books, 1992), 59–100; Henry Brandon, *The Retreat of American Power: The Inside Story of How Nixon and Kissinger Changed Foreign Policy for Years to Come* (New York: Doubleday, 1973), 218–46; Allen J. Matusow, *Nixon's Economy: Booms, Busts, Dollars and Votes* (Lawrence, KS: University Press of Kansas, 1998), 149–81; John S. Odell, *U.S International Monetary Policy: Markets, Power, and Ideas as Sources of Change* (Princeton, NJ: Princeton University Press, 1982), 165–340; and PhD dissertations by the following: Joanne Gowa, "Explaining Large Scale Policy Change: Closing the Gold Window,

1971" (Princeton University, 1980), Thomas Austin Forbord, "The Abandonment of Bretton Woods: The Political Economy of U.S. International Monetary Policy" (Harvard University, 1980), and Christen Thomas Ritter, "Closing the Gold Window: Gold, Dollars, and the Making of Nixonian Foreign Economic Policy" (University of Pennsylvania, 2007).

16   "If historians searched for the precise date": William Greider, *Secrets of the Temple: How the Federal Reserve Runs the Country* (New York: Simon and Schuster, 1987), 334.

Chapter 1: Richard Nixon Ascending

19   both chambers were controlled by Democrats: David S. Broder, "Nixon Wins With 290 Electoral Votes," *Washington Post*, November 7, 1968.

19   "Mr. Nixon starts with no clear mandate from the people": James Reston, "From Promise to Policy: Many Pitfalls Await Efforts by Nixon to Redeem Pledges and Unify Nation," *New York Times*, November 7, 1968.

20   "They were proclamations to be filled in": This quote and the following from Richard Reeves, *President Nixon: Alone in the White House* [e-book] (New York: Simon and Schuster, 2001), "Introduction."

20   As of the date he took office, nearly forty thousand American soldiers had died: Electronic Records Reference Report, Vietnam War U.S. Military Fatal Casualty Statistics, Military Records, National Archives, https://www.archives.gov/research/military/vietnam-war/casualty-statistics.

20   and two hundred more were being killed each week: Reeves, *President Nixon* [e-book], "Introduction."

22   the "New Isolationism": See, for example, James A. Johnson, "The New Generation of Isolationists," *Foreign Affairs* 49, no. 1 (October 1970): 136–46.

23   "New military programs [were] fiercely attacked": Kissinger, *White House Years*, 161.

23   trade and international finance appeared at the center of America's traditional foreign relations agenda: See Richard N. Cooper, "Trade Policy Is Foreign Policy," *Foreign Policy*, no. 9 (1972): 18–36.

24   "While we concerned ourselves with the NATO order of battle": Abraham Ribicoff, *Trade Policies in the 1970s: Report by Senator Abraham Ribicoff to the Committee on Finance, United States Senate*, S.Prt. 92-1 (Washington, DC: U.S. Government Printing Office, 1971), 11.

25 in comprehensive reports: See, for example, the president's "First Annual Report to the Congress on United States Foreign Policy for the 1970s, February 18, 1970," doc. 45 in *Public Papers of the Presidents of the United States: Richard Nixon, 1970* (Washington, DC: U.S. Government Printing Office, 1971).

Chapter 2: The Economic Crisis

28 amounting to about 8 percent in 1970: Calculated from United States President and Council of Economic Advisers, *Economic Report of the President (1972)* (Washington, DC: U.S. Government Printing Office, 1972), Table B-1, "National Income or Expenditure," 195, and Table B-87, "U.S. Balance of Payments," 296.

28 (For countries like West Germany and Japan): World Bank Group, "Trade (% of GDP)—Germany," chart, World Bank, https://data .worldbank.org/indicator/NE.TRD.GNFS.ZS?end=2018&locations= DE&start=1960&view=chart; World Bank Group, "Trade (% of GDP)—Japan," chart, World Bank, https://data.worldbank.org/ indicator/NE.TRD.GNFS.ZS?end=2018&locations=JP&start= 1960&view=chart.

28 By 1970, inflation had reached 5.5 percent: United States President and Council of Economic Advisers, *Economic Report of the President (1972)*, 41.

30 Nixon's budget deficit was $11 billion: United States President and Council of Economic Advisers, *Economic Report of the President (1971)* (Washington, DC: U.S. Government Printing Office, 1971), 23–27.

31 "performance of the economy disappointed many expectations and intentions": United States President and Council of Economic Advisers, *Economic Report of the President (1971)*, 28.

31 "Events during 1970 had justified these fears": United States Congress, "Report of the Joint Economic Committee, Congress of the United States, on the February 1971 Economic Report of the President" (Washington, DC: U.S. Government Printing Office, 1971), 37.

31 On Wednesday, November 18, 1970: Matusow, *Nixon's Economy*, 82.

31 "We are facing the greatest economic test of the postwar era": United States President and Council of Economic Advisers, *Economic Report of the President (1971)*, 3.

32 "If there is not a sustained pickup in the months ahead": Matusow, citing *Newsweek*, January 25, 1971, in his *Nixon's Economy*, 93.

32 "The first months of 1971": Richard M. Nixon, *RN: The Memoirs*

of Richard Nixon [e-book] (New York: Simon and Schuster, 2013), "1971."

32   Indeed, the consensus among economists and Wall Street: Terry Robards, "Elections Viewed Lacking as an Economic Mandate," *New York Times*, November 5, 1970.

Chapter 3: A Run on the Dollar?

38   The United States had run a surplus in its merchandise trade since 1893: Peter G. Peterson, "The United States in the Changing World Economy" (Washington, DC: U.S. Government Printing Office, 1971), 10.

39   "Such a deterioration might well trigger an acute confidence crisis": Gottfried Haberler, "Report of Task Force on U.S. Balance of Payments to the President-Elect" (Unpublished, January 1969), 14, RMN Presidential Library.

40   $10 billion in 1970 had increased to $30 billion in 1971: Matusow, *Nixon's Economy*, 143.

40   "I held to the hope that a major crisis could be avoided": Volcker and Gyohten, *Changing Fortunes*, 73.

41   "The anxieties about the huge balance of payments deficit": Brandon, *The Retreat of American Power*, 244.

41   In early May 1971, the long-feared crisis erupted: Sources for this section include: Robert Solomon, *The International Monetary System, 1945–1981* (New York: Harper and Row, 1982); Harold James, *International Monetary Cooperation Since Bretton Woods* (Washington, DC: International Monetary Fund and Oxford University Press, 1996); Charles A. Coombs, *The Arena of International Finance* (New York: Wiley, 1976); Luke A. Nichter, *Richard Nixon and Europe: The Reshaping of the Postwar Atlantic World* (Cambridge, UK, and New York: Cambridge University Press, 2017); plus contemporary press articles.

43   On May 12, New York senator Jacob Javits, a centrist Republican and key voice: "Gold Outglitters the West German Mark On World Markets; Speculators Sit Tight," *Wall Street Journal*, May 13, 1971.

43   On May 11, West German chancellor Willy Brandt wrote to Nixon: Nichter, *Richard Nixon and Europe*, 46.

43   "There is a widespread concern in Europe": Nichter, *Richard Nixon and Europe*, 47.

44   "Traders all over the world sensed a total breakdown": Coombs, *The Arena of International Finance*, 214.

Chapter 4: Richard M. Nixon

For this chapter, I relied on: Nixon, *RN*; Reeves, *President Nixon* [e-book]; Elizabeth Drew, *Richard M. Nixon* (New York: Times Books, 2007); Safire, *Before the Fall*; H. R. Haldeman, *The Haldeman Diaries: Inside the Nixon White House* (New York: G. P. Putnam's, 1994); Matusow, *Nixon's Economy*; Herbert Stein, *Presidential Economics: The Making of Economic Policy from Roosevelt to Reagan and Beyond* (New York: Simon and Schuster, 1984); extensive reporting from the *New York Times*, *Washington Post*, and *Wall Street Journal*; and personal interviews I conducted with Henry Kissinger, George Shultz, and Paul Volcker.

48   average age forty-six: Average based on their ages as of January 1, 1969.

49   "It is now evident that the President-elect": Roscoe Drummond, "Nixon Appointments Aim to Win Support of Center," *Washington Post*, December 18, 1968.

49   "This is where the votes are": Drummond, "Nixon Appointments Aim to Win Support of Center."

50   "before the fall": Safire, *Before the Fall*.

52   "It is the greatest comeback since Lazarus": James Reston, "A Remarkable Comeback for Nixon," *New York Times*, August 9, 1968.

53   Nixon gave a long answer: Richard Nixon, "Informal Remarks in Guam with Newsmen, July 25, 1969," doc. 279 in *Public Papers of the Presidents of the United States: Richard Nixon, 1969* (Washington, DC: U.S. Government Printing Office, 1971); see also Nixon, *RN*, 394.

53   Years later, Henry Kissinger recalled: Kissinger, *White House Years*, 225.

56   "The [Nixon] Doctrine set a tone": This quote and the following are from Brandon, *The Retreat of American Power*, 81.

57   "I've always thought the country could run itself": Reeves, *President Nixon* [e-book], chapter 2.

57   "felt that for Republicans, economics was not a winning issue": Stein, *Presidential Economics*, 138.

57   "Nixon repeatedly interrupted Cabinet meetings": Reeves, *President Nixon* [e-book], chapter 11.

57   "When you start talking about inflation in the abstract": Matusow, *Nixon's Economy*, 16.

57   "But when unemployment goes up one half of one percent": Matusow, *Nixon's Economy*, 16.

57   "I do not go along with the suggestion": Richard M. Nixon, "The

President's News Conference of January 27, 1969," doc. 10 in *Public Papers of the Presidents of the United States: Richard Nixon, 1969.*

57 "My first job in government was with the old Office of Price Administration": Richard M. Nixon, "Address to the Nation on the Rising Cost of Living, October 17, 1969," doc. 395 in *Public Papers of the Presidents of the United States: Richard Nixon, 1969.*

58 "A question I am increasingly asked": Paul A. Samuelson, "Gold," *Newsweek*, October 14, 1968.

58 "President-elect Nixon is going to find himself": Paul A. Samuelson, "The New Economics," *Newsweek*, November 25, 1968.

58 "I do not want to be bothered with international monetary matters": "Editorial Note, March 2, 1970 (Nixon Memo to Haldeman on International Economic Policy)," doc. 38 in *Foreign Relations of the United States, 1969–1976*, vol. 3, *Foreign Economic Policy; International Monetary Policy, 1969–1972* (Washington, DC: U.S. Government Printing Office, 2001).

59 "I would like to defer": Edwin L. Dale Jr., "David Kennedy Hedges on $35 Gold," *New York Times*, December 18, 1968.

59 "We do not anticipate any changes in the price of gold": Hobart Rowen, "Fear of Rising Gold Price Persists Despite Disclaimer," *Washington Post*, December 19, 1968.

59 Nixon believed that gold was an anachronism in monetary affairs: Matusow, *Nixon's Economy*, 126.

59 "Now is the time to examine our international monetary system": Richard M. Nixon, "Remarks to Top Personnel at the Department of the Treasury, February 14, 1969," doc. 49 in *Public Papers of the Presidents of the United States: Richard Nixon, 1969.*

60 "I believe that the whole world will be served by moving toward freer trade": Richard M. Nixon, "The President's News Conference of February 6, 1969," doc. 34 in *Public Papers of the Presidents of the United States: Richard Nixon, 1969.*

60 "As far as the textile situation is concerned": Nixon, "The President's News Conference of February 6, 1969."

60 he told the cabinet: "Information Memorandum from C. Fred Bergsten of the National Security Council Staff to the President's Assistant for National Security Affairs (Kissinger), April 14, 1969," doc. 19 in *Foreign Relations of the United States, 1969–1976*, vol. 3, *Foreign Economic Policy; International Monetary Policy, 1969–1972* (Washington, DC: U.S. Government Printing Office, 2001).

60   social programs increased as a percentage of overall government spending from 33 percent to 50 percent: Matusow, *Nixon's Economy*, 35.

60   Speechwriter William Safire said: This sentence and the following description are from Safire, *Before the Fall*, 599.

60   Biographer Richard Reeves tells of how Arthur Burns: Description and quote from Reeves, *President Nixon* [e-book], chapter 3.

61   "I have made some bad decisions": Safire, *Before the Fall*, 103.

61   Henry Kissinger once said that Nixon had a motto: Quote and Kissinger's explanation from Winston Lord, *Kissinger on Kissinger: Reflections on Diplomacy, Grand Strategy, and Leadership* [e-book] (New York: All Points Books, 2019), chapter 1.

61   He felt ill at ease with almost everyone but his wife and daughters: R. W. Apple Jr., "The 37th President; Richard Nixon, 81, Dies; A Master of Politics Undone by Watergate," *New York Times*, April 23, 1994.

61   "It's been a year": Quote and description of the scene comes from Safire, *Before the Fall*, 603.

61   "Great men of action": Reeves, *President Nixon* [e-book], chapter 1.

62   "[Nixon] wanted to seem slightly out of reach": Safire, *Before the Fall*, 619.

62   "I just get up every morning to confound my enemies": Reeves, *President Nixon* [e-book], "Introduction."

62   "Richard Nixon went up the walls of life with his claws": Quoted in John Herbers, "The 37th President; In Three Decades, Nixon Tasted Crisis and Defeat, Victory, Ruin and Revival," *New York Times,* April 24, 1994.

62   "The extent to which [the president] retained command": Brandon, *The Retreat of American Power*, 63.

62   "Nixon's style is that of a judge": Juan Cameron, "Richard Nixon's Very Personal White House," *Fortune*, July 1970. All quotations in this paragraph.

Chapter 5: John B. Connally Jr.

For this chapter on Connally, I benefited enormously from interviews with men who worked closely with Connally or knew him during his Texas years: Larry Temple (January 24, 2018), Ben Barnes (January 12, 2018), Julian Read (January 24, 2018), James Baker (November 28, 2017). See page 421 for more details on the interviews. I also drew on interviews with others who worked directly with him when he was secretary of the treasury: C. Fred Bergsten (October 5, 2017), John Petty (October 11 and November 9, 2017), Michael Bradfield (July 12,

2017), Paul Volcker (July 26, 2017), and Robert Hormats (July 12 and December 11, 2017). Also, I relied on James Reston Jr.'s biography of Connally, *The Lone Star: The Life of John Connally* (New York: Harper and Row, 1989), and Connally's memoir, *In History's Shadow: An American Odyssey* (New York: Hyperion, 1993).

63   A moderate Republican, he was even considered for the top Treasury post: Hobart Rowen, "Profiles of the Nixon Cabinet: David M. Kennedy," *Washington Post*, December 12, 1968.

63   What seemed to matter most to Nixon: Nixon, *RN*, 339.

64   at virtually every critical moment, Nixon told Haldeman to check with Connally: Stephen E. Ambrose and Douglas Brinkley, *Rise to Globalism: American Foreign Policy Since 1938* (New York: Penguin Books, 1997), 3.

66   "Every Cabinet should have at least one potential President in it": Safire, *Before the Fall*, 498.

67   "When he walked into a room he radiated ambition": Peter G. Peterson, *The Education of an American Dreamer: How a Son of Greek Immigrants Learned His Way from a Nebraska Diner to Washington, Wall Street, and Beyond* [e-book] (New York: Hachette Book Group, 2009), chapter 7.

67   "Connally meant stimulation, excitement, political savvy to Nixon": Safire, *Before the Fall*, 497.

67   "Connally's swaggering self-assurance": Kissinger, *White House Years*, 951.

67   "It was the most impressive presentation on Vietnam": Author's interview of Larry Temple, January 24, 2018.

68   Connally recollected to the aide a host of numbers in that speech: Description of this incident drawn from Paul Burka, "The Truth About John Connally," *Texas Monthly*, November 1979.

69   "Brains not brawn": This quote and the following are from Connally, *In History's Shadow*, 221.

69   "You operate like I do . . . We're termites": Reston, *The Lone Star*, 175.

69   "ambitious, political animals who came out of the same jungle": Reston, *The Lone Star*, 302.

70   "I can use raw power as well as anyone": Reston, *The Lone Star*, 418.

70   "Much of what Martin Luther King said": Reston, *The Lone Star*, 343.

70   "It's not enough for Connally to beat you": Burka, "The Truth About John Connally."

70   "He's got to rub your nose in the dirt": Burka, "The Truth About John Connally."

70   "You will be measured in this town [Washington] by the enemies you destroy": Kissinger, *White House Years*, 951.

71   He went on to describe Connally's qualifications: Richard M. Nixon, "Remarks on Plans to Nominate Secretary Kennedy as Ambassador-at-Large and Governor Connally as Secretary of the Treasury, December 14, 1970," doc. 460 in *Public Papers of the Presidents of the United States: Richard Nixon, 1970*.

71   "He had repaired his relations with a hostile Congress": Richard Whalen, "The Nixon-Connally Arrangement," *Harper's*, August 1971.

72   "I say let's run the risk": Safire, *Before the Fall*, 504.

72   "If you lose, you lose big": Safire, *Before the Fall*, 504.

73   This made Connally furious: Author's interview of Robert Hormats, December 11, 2017.

73   Henry Brandon wrote that Connally couldn't care less: Brandon, *The Retreat of American Power*, 229.

73   He had no attachment to the Bretton Woods structures: Gowa, "Explaining Large Scale Policy Change," 188–89.

74   "Do you believe in the divine creation": Account of this exchange taken from "Connally's Hard Sell Against Inflation," *BusinessWeek*, no. 2184 (July 10, 1971).

74   "I can play it round or I can play it flat": Reston, *The Lone Star*, 403.

75   those of West Germany increased 200 percent and those of Japan rose by 400 percent: Peterson, "The United States in the Changing World Economy," 8.

75   its exports grew by 17 percent per year in the 1960s: Figures from Robert Scalapino, *American-Japanese Relations in a Changing Era* (Washington, DC: Georgetown University, Center for Strategic and International Studies, 1972), 20.

75   Taking the European Community and Japan together: U.S. Commission on International Trade and Investment Policy (Williams Commission), *United States International Economic Policy in an Interdependent World; Report to the President* (Washington, DC: U.S. Government Printing Office, 1971), 65.

76   "I'll always back you, but if I don't get your loyalty": Author's interview with Michael Bradfield, former deputy counsel at the U.S. Treasury, July 12, 2017.

76   "When I took over as Secretary of the Treasury": Connally, *In History's Shadow*, 235.

77 Yet, "[a]fter a tutorial or two": Author's interview of John Petty, November 9, 2017.

77 "I remember my sense of relief": This quote and the following from Volcker and Gyohten, *Changing Fortunes*, 72.

77 Connally said he couldn't wait, grabbed it out of Volcker's fist, and left: Preceding account of Volcker and Connally's interaction from author's interview with Paul Volcker, July 26, 2017.

77 "I want to screw the foreigners before they screw us": Author's interview of C. Fred Bergsten, October 5, 2017. Bergsten was present at the meeting.

Chapter 6: Paul A. Volcker Jr.

For background on Volcker, I relied heavily on William L. Silber, *Volcker: The Triumph of Persistence* (New York: Bloomsbury Press, 2012); Volcker and Gyohten, *Changing Fortunes*; Interview with Paul A. Volcker, January 28, 2008–March 24, 2010, Federal Reserve Board Oral History Project, https://www.federalreserve.gov/aboutthefed /files/paul-a-volcker-interview-20080225.pdf [hereafter: "Volcker Oral History Project"]; and Paul Volcker and Christine Harper, *Keeping At It: The Quest for Sound Money and Good Government* (New York: PublicAffairs, 2018). I also benefited greatly from my interviews with Volcker himself (July 26, 2017), Michael Bradfield (July 12, 2017), John Petty (October 11 and November 9, 2017), Robert Hormats (July 12 and December 11, 2017), and Fred Bergsten (October 5, 2017).

78 "he resembled a civil servant from the British system": Greider, *Secrets of the Temple*, 68.

79 He smoked cheap cigars: Neil Irwin, "Paul A. Volcker, Fed Chairman Who Curbed Inflation by Raising Interest Rates, Dies at 92," *Washington Post*, December 9, 2019.

79 Connally once threatened to fire him: Greider, *Secrets of the Temple*, 68.

82 Shortly after he assumed office on Monday, January 20, 1969: Description of events in this paragraph are drawn from Silber, *Volcker*, 55; and Volcker and Gyohten, *Changing Fortunes*, 64.

83 Volcker was an enigma: Author's interviews of Michael Bradfield (July 12, 2017) and John Petty (October 11 and November 9, 2017).

83 "[He] mulled things over in his own mind": Gowa, "Explaining Large Scale Policy Change," 110.

83 "the $35 gold price was sacred": Volcker Oral History Project, 49.

83   That action, Volcker said, "was anathema to me": Volcker Oral History Project, 51.

83   "Price stability belongs to the social contract": Silber, *Volcker*, 53.

84   "We give government the right to print money": Silber, *Volcker*, 53.

86   a forty-eight-page, double-spaced document: See "Memorandum from Secretary of the Treasury Kennedy to President Nixon, June 23, 1969 (Includes Paper 'Basic Options in International Monetary Affairs,' by Paul Volcker)," doc. 130 in *Foreign Relations of the United States, 1969–1976*, vol. 3, *Foreign Economic Policy; International Monetary Policy, 1969–1972* (Washington, DC: U.S. Government Printing Office, 2001).

87   Volcker did the presentation using a series of flip charts: Silber, *Volcker*, 62.

90   it was seen as the ultimate weapon: Gowa, "Explaining Large Scale Policy Change," 289–90.

91   "Good job, and keep me posted on where we stand": Silber, *Volcker*, 67.

92   Volcker replied: Silber, *Volcker*, 67.

92   "[w]as included among the other options, as far as I was concerned": Volcker and Gyohten, *Changing Fortunes*, 67.

92   He later acknowledged: Forbord, "The Abandonment of Bretton Woods," 107.

93   He felt that the treasury undersecretary cared more about international economic relations: Silber, *Volcker*, 84–85.

Chapter 7: Arthur F. Burns

For background on Arthur Burns, I drew heavily on Arthur F. Burns, *Inside the Nixon Administration: The Secret Diary of Arthur Burns, 1969–1974*, edited by Robert H. Ferrell (Lawrence, KS: University of Kansas Press, 2010); Wyatt C. Wells, *Economist in an Uncertain World: Arthur F. Burns and the Federal Reserve, 1970–78* (New York: Columbia University Press, 1994); Arthur F. Burns, *Reflections of an Economic Policy Maker: Speeches, Congressional Statements: 1969–1978* (Washington, DC: American Enterprise Institute, 1978); Stephen H. Axilrod, *Inside the Fed: Monetary Policy and Its Management, Martin Through Greenspan to Bernanke* (Cambridge, MA: The MIT Press, 2011); Matusow, *Nixon's Economy*; Stein, *Presidential Economics*; Leonard S. Silk, *Nixonomics: How the Dismal Science of Free Enterprise Became the Black Art of Controls* (New York: Praeger Publishers, 1972); Nixon, *RN*; James L. Pierce, "The Political Economy of Arthur Burns," *The Journal of Finance* 34, no. 2 (1979): 485–96; Burton A. Abrams, "How Richard Nixon Pressured

Arthur Burns: Evidence from the Nixon Tapes," *Journal of Economic Perspectives* 20, no. 4 (2006): 177–88.

96 "He would clean, fill, light, relight, empty and refill his pipe": Wells, *Economist in an Uncertain World*, 7.

97 "[Burns] takes so long to answer a question between puffs": John Pierson, "White House Power: Arthur Burns Provides Conservative Influence on Domestic Programs," *Wall Street Journal*, May 20, 1969.

97 "His brand of humor is described as 'dry' or 'non-existent'": "How Burns Will Change the Fed," *BusinessWeek*, October 25, 1969.

97 "I've never heard anyone mention diplomacy, tact, humility": "How Burns Will Change the Fed."

97 "By the time Burns finished, the staff was choking with laughter": Wells, *Economist in an Uncertain World*, 45.

97 He often acted as mentor: Wells, *Economist in an Uncertain World*, 49.

98 A few days later, the president took the occasion to talk to Burns: This account drawn from John Ehrlichman, *Witness to Power: The Nixon Years* (New York: Simon and Schuster, 1982), 248–49.

98 My relations with the Fed will be different: This exchange between Nixon and Burns taken from Ehrlichman, *Witness to Power*, 248–49.

101 not terribly interested in theoretical analysis: Axilrod, *Inside the Fed*, 58.

101 "The argument between the Friedmanites and the Keynesians": Quoted in Lawrence A. Malkin, "A Practical Politician at the Fed," *Fortune* 83 (May 1971), 260.

102 In the spring of 1971, however, he met with the French minister of economy and finance: Burns, *Inside the Nixon Administration*, 42 (May 22, 1971).

103 "Eisenhower liked to talk about the independence of the Federal Reserve": Matusow, *Nixon's Economy*, 20.

103 he had this exchange with Sen. William Proxmire: The following exchange taken from U.S. Senate, Committee on Banking and Currency, and Ninety-First Congress, "Nomination of Arthur F. Burns: Hearing Before the Committee on Banking and Currency" (December 18, 1969), 5.

104 "I am convinced the President would do anything to be reelected": Burns, *Inside the Nixon Administration*, 37 (Monday, March 8, 1971).

104 "The harassing of the Fed by the President and his pusillanimous staff": Burns, *Inside the Nixon Administration*, 37 (Monday, March 8, 1971).

104 "that his friendship is one of the three that has counted most in my life": Burns, *Inside the Nixon Administration*, 39 (Sunday, March 21, 1971).

105 Burns confided to his diary of "the brutality of Nixon's language":
All quotations in this paragraph from Burns, *Inside the Nixon Administration*, 44–48 (Thursday, July 8, 1971).

105 Burns's desire to be Nixon's friend: Burns, *Inside the Nixon Administration*, 28 (November 23, 1970).

105 "pathetic slob": Burns, *Inside the Nixon Administration*, 5.

105 "obsessed with his own status": Burns, *Inside the Nixon Administration*, 8.

105 "amoral leanings": Burns, *Inside the Nixon Administration*, 33.

105 "pernicious and stultifying force": Burns, *Inside the Nixon Administration*, 37.

105 "loyal and devoted servant": Burns, *Inside the Nixon Administration*, 121.

105 John Connally lacked capacity for leadership: Burns, *Inside the Nixon Administration*, 43.

105 "egomaniacal approach": Burns, *Inside the Nixon Administration*, 124.

105 "an indecisive man full of flaws and anxieties": Burns, *Inside the Nixon Administration*, 62.

106 "You see, Dr. Burns," Nixon said to the assembled crowd: Richard M. Nixon, "Remarks at the Swearing in of Dr. Arthur F. Burns as Chairman of the Board of Governors of the Federal Reserve System, January 31, 1970," doc. 21 in *Public Papers of the Presidents of the United States: Richard Nixon, 1970*.

106 "I hope that independently he will conclude that my views are the ones that should be followed": Nixon, "Remarks at the Swearing in of Dr. Arthur F. Burns as Chairman of the Board of Governors of the Federal Reserve System, January 31, 1970."

106 "My duties at the Federal Reserve Board, I think, can be described in one sentence": Arthur Burns, "Remarks of the President and Dr. Arthur F. Burns at the Swearing in of Dr. Burns as Chairman, January 31, 1970," *Weekly Compilation of Presidential Documents* 6, no. 5 (Washington, DC: U.S. Government Printing Office, 1970), 98.

107 "Responsibility for recession is directly on the Fed": from Ehrlichman, *Witness to Power*, 250.

107 "Arthur has a way of holding the money supply hostage": Matusow, *Nixon's Economy*, 62.

107 "this Fed won't be independent if it's the only thing I do": Matusow, *Nixon's Economy*, 62.

108 He started with suggestions for voluntary measures: Arthur F. Burns, "Inflation: The Fundamental Challenge to Stabilization Policies: Remarks before the Seventeenth Annual Monetary Conference of the

American Bankers Association, Hot Springs, Virginia, May 18, 1970," in his *Reflections of an Economic Policy Maker*, 91–102.

108 Several members of Nixon's cabinet, within the confines of internal deliberations, favored them: Matusow, *Nixon's Economy*, 67.

108 Paul McCracken, who had publicly opposed controls: Erwin C. Hargrove and Samuel A. Morley, eds., *The President and the Council of Economic Advisers: Interviews with CEA Chairmen* (Boulder, CO: Westview Press, 1984), 316.

108 the Business Council, the most prestigious group of CEOs, pushed voluntary controls: Albert R. Hunt, "Leaders' Lament: Business Council's Inflation Gripes May Renew Nixon Leadership Issue," *Wall Street Journal*, October 19, 1970.

108 It was designed to embarrass the president: Author's interview with Ken McLean, October 6, 2017, and Matusow, *Nixon's Economy*, 67.

109 "Burns will get it in the chops": Ehrlichman, *Witness to Power*, 250.

109 "I emphasized time was short": Burns, *Inside the Nixon Administration*, 29 (November 20, 1970).

109 On Friday, December 4, in a speech at the Waldorf Astoria: Richard M. Nixon, "Remarks at the Annual Meeting of the National Association of Manufacturers, December 4, 1970," doc. 447 in *Public Papers of the Presidents of the United States: Richard Nixon, 1970*.

109 Burns gave another speech a few days later, taking an even tougher stand: Arthur F. Burns, "The Basis for Lasting Prosperity, Address by Arthur F. Burns in the Pepperdine College Great Issues Series, December 7, 1970," in his *Reflections of an Economic Policy Maker*, 103–15.

110 the "Accord of 1970": Leonard S. Silk, "The Accord of 1970," *New York Times*, December 9, 1970.

110 "In effect, Burns was offering the White House a concordant": "Nixon Is Shifting to a Harder-Hitting Game," *BusinessWeek*, no. 2154 (December 12, 1970).

110 "I do not intend to impose wage and price controls": United States President and Council of Economic Advisers, *Economic Report of the President (1971)*, 7.

110 "Mandatory price and wage controls are undesirable": United States President and Council of Economic Advisers, *Economic Report of the President (1971)*, 80.

110 "The administration and the economy were engaged in a race": Stein, *Presidential Economics*, 156.

110 "The question was whether the administration's disinflation program":
Stein, *Presidential Economics*, 156.

Chapter 8: George P. Shultz

For background on Shultz, I relied heavily on George P. Shultz,
Turmoil and Triumph: My Years as Secretary of State [e-book] (New
York: Scribner's, 1993); George P. Shultz, *Learning from Experience*
(Stanford, CA: Hoover Institution Press, 2016); Matusow, *Nixon's
Economy*; "Hearing Before the Committee on Labor and Public Welfare
on George P. Shultz to Be Secretary of Labor (1969)"; a number of
biographical portraits in the *New York Times*, *Wall Street Journal*,
and *BusinessWeek*; and my personal interviews with him (August 16,
2017) and with his two deputies, Arnold Weber (August 14, 2017) and
Kenneth Dam (July 18, 2017).

113 "I don't think that the President looks at me as a great font of wisdom":
This quote and following from James M. Naughton, "Shultz Quietly
Builds Up Power in Domestic Field," *New York Times*, May 31, 1971.
114 From that perch, Shultz would meet with Nixon once or twice each day:
"The Architect of Nixon's New Economics," *BusinessWeek*, March 20, 1971.
114 He took over the 7:30 a.m. White House meeting: John Pierson,
"Who's in Charge? Coming Cabinet Moves Point Up the Big Decline
of Secretaries' Role," *Wall Street Journal*, November 20, 1970.
114 "He is one of the President's three or four top confidants": "The
Architect of Nixon's New Economics."
114 "Every White House eventually produces an individual who is relied
upon so heavily": Naughton, "Shultz Quietly Builds Up Power in
Domestic Field."
114 "George Shultz has served with distinction": Bruce Winters, "Shultz Noted
as a Skilled, Cool Labor Secretary," *The Baltimore Sun*, June 11, 1970.
115 "a shirt-sleeve type, pipe-smoking, and somewhat rumpled": "The
Architect of Nixon's New Economics."
115 He spoke softly, raising his voice: "The Architect of Nixon's New
Economics."
115 "Time and again he would work with almost inhuman patience":
Volcker and Gyohten, *Changing Fortunes*, 118.
115 "With the changing political winds": Matusow, *Nixon's Economy*, 29.
115 "He who walks in the middle": Safire, *Before the Fall*, 265.

115 Shultz impressed Nixon with his ability to translate ideas into action: Stein, *Presidential Economics*, 145.

115 "the only real knowledgeable economist": Haldeman, *Haldeman Diaries*, 70 (February 12–15, 1971).

115 Safire judged Shultz the fiercest advocate of the free-market economy: Safire, *Before the Fall*, 491.

115 The key to his ideological beliefs was his strong association with the "Chicago School": See Johan Van Overtveldt, *The Chicago School: How the University of Chicago Assembled the Thinkers Who Revolutionized Economics and Business* (Chicago: Agate, 2007), "Introduction," 1–17.

116 "George is a man of principle, but he is not an ideologue": This quote and the following from "The Architect of Nixon's New Economics."

117 As for Connally: Author's interview with George Shultz, August 16, 2017. Shultz's views on Connally and Kissinger that were expressed during the interview.

117 "stiff as a board": Matusow, *Nixon's Economy*, 27.

117 "the grayest man in a gray cabinet": "The Architect of Nixon's New Economics."

117 Shultz gave one speech that he himself would cite time and again: In my interview of him in the summer of 2017, almost a half century later, he emphasized this speech to me and gave me a copy of a new presentation he was planning to deliver in which he also referred to the original speech.

117 The speech was called "Prescription for Economic Policy—'Steady as You Go': George P. Shultz, "Prescription for Economic Policy: 'Steady As You Go,'" Address Before the Economic Club of Chicago, April 22, 1971, https://web.stanford.edu/~johntayl/Shultz%20on%20Steady%20As%20You%20Go.pdf.

Chapter 9: Peter G. Peterson

As background for this chapter I used Ken Auletta, *Greed and Glory on Wall Street: The Fall of the House of Lehman* (New York: Random House, 1986), 35–37; Peterson, *Education of an American Dreamer*; various obituaries, including Robert D. Hershey Jr., "Peter G. Peterson, a Power from Wall St. to Washington, Dies at 91," *New York Times*, March 20, 2018; extensive articles on Peterson in *BusinessWeek* and *Fortune*; and interviews of his contemporaries in the Nixon administration.

122 "Nixon was envisioning the interconnected geo-economic world": Peterson, *Education of an American Dreamer*, chapter 7, "Washington: Round One."

122 "He is a man that has been described by his colleagues": Richard M. Nixon, "Remarks Announcing Appointment of Peter G. Peterson as Assistant to the President for International Economic Affairs and Executive Director, Council on International Economic Policy, January 19, 1971," doc. 19 in *Public Papers of the Presidents of the United States: Richard Nixon, 1971* (Washington, DC: U.S. Government Printing Office, 1972).

122 "It was the surest sign so far that I had left the area of quantifiable data": Peterson, *Education of an American Dreamer*, chapter 7, "Washington: Round One."

122 "Pete Peterson is stepping into one of the toughest jobs in Washington": "Mr. Peterson's Assignment," *Fortune*, March 1971.

123 "had a mind full of penetrating questions": Safire, *Before the Fall*, 497.

124 (John Petty, one of Connally's key assistant secretaries, told me in an interview): Author's interviews of John Petty, October 11 and November 9, 2017.

124 The report made several key points: For the following section, I used a confidential version of the report, Peter G. Peterson, "The United States in the Changing World Economy (Confidential Version)," April 1971, Box 108477, Papers of Paul A. Volcker, New York Federal Reserve Archive; for the public version, see Peterson, "The United States in the Changing World Economy," and a description in Matusow, *Nixon's Economy*, 132–39.

125 the report went to Nixon, who was so taken: Matusow, *Nixon's Economy*, 137, referring to Ehrlichman notes taken on July 27, 1971. The following two quotations are from the same source.

126 "It's terribly important that we be #1 economically": Matusow, *Nixon's Economy*, 136.

126 In late June, Nixon and Connally discussed: Matusow, *Nixon's Economy*, 136, citing Haldeman, *Haldeman Diaries* (June 27, 1971).

126 "Mr. Peterson's graphs and charts": John Pierson, "Trade Tightrope: Nixon Aide Peterson Has Controversial Ideas on Overseas Dealings," *Wall Street Journal*, July 6, 1971.

126 John Connally was a particular booster of Peterson's report: Matusow, *Nixon's Economy*, 133.

126 "the starting point for dramatic changes in economic policy": Juan

Cameron, "The Last Reel of 'Mr. Peterson Goes to Washington,'" *Fortune*, March 1973.

126 On Monday, July 12, 1971, Peterson sent a confidential memo to Nixon: Peter G. Peterson, "Memorandum for the President: Projecting the Future Development of the U.S," July 12, 1971, CIEP Folder, National Archives at College Park, College Park, MD.

127 "We have experienced our first smell of defeat": Peterson, "Memorandum for the President," July 12, 1971.

127 "We need to sense and shape a new and exciting future": Peterson, "Memorandum for the President," July 12, 1971.

127 "This is where we are . . . this is where we want to be": Peterson, "Memorandum for the President," July 12, 1971.

128 "I naively assumed that Connally would have been supportive": Peterson, *Education of an American Dreamer*, chapter 7, "Washington: Round One."

128 "Connally had reduced Peterson to the role of spectator": Kissinger, *White House Years*, 952.

128 He dined with people like Katharine Graham: Auletta, *Greed and Glory on Wall Street*, 38.

129 "overzealous in seeking public credit for his accomplishments": Cameron, "The Last Reel of 'Mr. Peterson Goes to Washington.'" In addition to the Cameron article, these ideas also came out in my interviews with Arnold Weber, associate director for the Office of Management and Budget under George Shultz, August 14, 2017, and John Petty, October 11 and November 9, 2017.

129 "I don't give a damn what it takes": Peterson, *Education of an American Dreamer*, chapter 7, "Washington: Round One."

Chapter 10: Other Players— Paul W. McCracken and Henry A. Kissinger

For this chapter, I relied on several books, including the following. For Paul McCracken: Sidney L. Jones, *Public and Private Economic Advisor: Paul W. McCracken* (Lanham, MD: University Press of America, 2000); Stein, *Presidential Economics*; and Hargrove and Morley, eds., *The President and the Council of Economic Advisers: Interviews with CEA Chairmen*. For Henry Kissinger: Kissinger, *White House Years*; Walter Isaacson, *Kissinger: A Biography* (New York: Simon and Schuster, 1992); Robert Dallek, *Nixon and Kissinger:*

Partners in Power (New York: HarperCollins, 2007); and David J. Rothkopf, *Running the World: The Inside Story of the National Security Council and the Architects of American Power* (New York: Public Affairs, 2005).

132 "The age of the superpowers is now drawing to an end": Quoted in Kermit Gordon, ed., *Agenda for the Nation: Papers on Domestic and Foreign Policy Issues* (Washington, DC: Brookings Institution, 1968), 597.

132 "And there must be a conviction that the United States cannot or will not carry all the burdens alone": Gordon, ed., *Agenda for the Nation*, 597.

132 "There was no doubt about how foreign policy was being made in the Nixon administration": Rothkopf, *Running the World*, 45.

135 Kissinger was deeply uncomfortable talking in a group on subjects that he could not dominate: Author's interview of C. Fred Bergsten, October 5, 2017.

Chapter 11: The Wolf at the Door

139 "We can't continue to hold a military, economic and political umbrella over the free world": Don Oberdorfer and Frank C. Porter, "Connally Urges Tough Trade Stance," *Washington Post*, April 26, 1971.

140 "If that's the way they feel, the United States should pull its Sixth Fleet out of the Mediterranean": Oberdorfer and Porter, "Connally Urges Tough Trade Stance."

140 "be willing to bear disproportionate economic costs does not fit the facts of today": John B. Connally, "Mutual Responsibility for Maintaining a Stable Monetary System, Address by John B. Connally Before the American Bankers Association at Munich, Germany on May 28," in *Department of State Bulletin* LXV, no. 1672 (Washington, DC: U.S. Government Printing Office, 1971), 45.

140 "The Nixon administration is dedicated to assuring the integrity and maintaining the strength of the dollar": Connally, "Mutual Responsibility for Maintaining a Stable Monetary System," 46.

140 "We are not going to change the price of gold": Connally, "Mutual Responsibility for Maintaining a Stable Monetary System."

141 To some, Connally proclaimed the Nixon Doctrine in tough, blunt terms: Nichter, *Richard Nixon and Europe*, 47.

141 To others, his remarks portended an inward, protectionist path:

Hobart Rowen, "U.S. Going Protectionist Following Monetary Crisis," *Washington Post*, June 7, 1971.

141   "the old monetary system was dead": Clyde H. Farnsworth, "Dollar Challenged," *New York Times*, May 28, 1971.

141   "That's my unalterable position today": Volcker and Gyohten, *Changing Fortunes*, 75.

141   Volcker's reservations about Connally's commitment: The following section is heavily based on: Volcker and Gyohten, *Changing Fortunes*; Volcker and Harper, *Keeping at It*; Silber, *Volcker*; Volcker Oral History Project; John S. Odell, "Going Off Gold and Forcing Dollar Depreciation," chapter 4 in his *U.S International Monetary Policy*; "Paper Prepared in the Department of the Treasury: 'Contingency,' May 8, 1971," doc. 152 in *Foreign Relations of the United States, 1969–1976*, vol. 3, *Foreign Economic Policy; International Monetary Policy, 1969–1972* (Washington, DC: U.S. Government Printing Office, 2001); Papers of Paul A. Volcker, Box 108477, New York Federal Reserve Archive; author's interview of Paul Volcker, July 26, 2017; author's interview of John Petty, October 11 and November 9, 2017; and author's interview of Michael Bradfield, July 12, 2017.

142   "I came to believe that sooner or later we would have to suspend our promise to convert dollars into gold": Volcker and Harper, *Keeping at It*, 67.

142   "We needed to find the right time to take the initiative": Volcker and Harper, *Keeping at It*, 67.

143   "Pressures on the international monetary system are rapidly building": "Paper Prepared in the Department of the Treasury: 'Contingency,' May 8, 1971."

143   "Reasonably foreseeable events": "Paper Prepared in the Department of the Treasury: 'Contingency,' May 8, 1971."

145   Volcker said it felt too protectionist for him: Joanne Gowa, *Closing the Gold Window: Domestic Politics and the End of Bretton Woods* (Ithaca, NY: Cornell University Press, 1983), 190; Volcker and Gyohten, *Changing Fortunes*, 75; author's interview of Paul Volcker, July 26, 2017.

145   "Embarking on this course": "Paper Prepared in the Department of the Treasury: 'Contingency,' May 8, 1971."

145   In a memo to the president: Paul McCracken, "Memorandum from the Chairman of the Council of Economic Advisers (McCracken) to President Nixon, June 2, 1971," doc. 157 in *Foreign Relations of the United States, 1969–1976*, vol. 3, *Foreign Economic Policy; International*

Monetary Policy, 1969–1972 (Washington, DC: U.S. Government Printing Office, 2001).

146 "With no joy, I have concluded that we must even be prepared": Paul McCracken, "Memorandum for the President, Subject: Quadriad Meeting—Monday, June 14, 1971," June 14, 1971, FRASER (Federal Reserve System digital library), https://fraser.stlouisfed.org/archival /1173/item/3391.

146 During the weekend of Saturday, June 26, 1971, the president and his economic advisors met at Camp David: Volcker and Gyohten, *Changing Fortunes*, 75.

146 In his diary, Haldeman paraphrased the president: Haldeman, *Haldeman Diaries*, 307.

146 Shultz was not happy: Ehrlichman, *Witness to Power*, 260.

146 "Just tell Arthur to report to Connally": Ehrlichman, *Witness to Power*, 260.

146 Connally was then told to brief the press: John B. Connally, "Remarks of Secretary of the Treasury John B. Connolly, Jr., at a Press Conference Following a Series of Discussions by Administration Officials on Economic and Budget Matters, June 29, 1971," in *Weekly Compilation of Presidential Documents* 7, no. 27 (Washington, DC: U.S. Government Printing Office, 1971), 1002–4.

147 they would just take time: Silk, *Nixonomics*, 17.

147 "death watch for Bretton Woods": Matusow, *Nixon's Economy*, 144.

147 "I found that for every one who expressed support for that foreign policy initiative": Nixon, *RN*, 518.

147 "If we don't propose a responsible new [economic] program": Nixon, *RN*, 518.

147 a Harris Poll conducted in July showed that 73 percent of respondents had an unfavorable view: This and Gallup Poll numbers from Ritter, "Closing the Gold Window," 234.

148 "the rules of economics are not working quite the way they used to": Arthur F. Burns, "Statement Before the Joint Committee of the U.S. Congress, July 23, 1971 ('The Economy in Mid-1971')," in his *Reflections of an Economic Policy Maker*, 117–27.

149 "It was like an explorer discovering": Silk, *Nixonomics*, 57.

149 "This week the deep hostility between the Administration and the chairman of the Fed broke out in the open": "The Administration Takes on the Fed," *BusinessWeek*, July 31, 1971.

149 "Clearly, officials are trying to force Burns back in line": "The Administration Takes on the Fed."

149 "[It] could become a confrontation of historic proportions": Richard F. Janssen and Albert R. Hunt, "War of Nerves Quickens: White House Hints It Plans Attack on Reserve Board's Independence," *Wall Street Journal*, July 29, 1971.

150 In a wide-ranging press conference: Richard M. Nixon, "The President's News Conference of August 4, 1971," doc. 250 in *Public Papers of the Presidents of the United States: Richard Nixon, 1971*, 849–61.

150 Burns wrote in his diary: Burns, *Inside the Nixon Administration*, 50 (August 22, 1971).

150 "The United States, as compared with that position we found ourselves in": Richard M. Nixon, "Remarks to Midwestern News Media Executives Attending a Briefing on Domestic Policy in Kansas City, Missouri, July 6, 1971," doc. 222 in *Public Papers of the Presidents of the United States: Richard Nixon, 1971*.

150 "But now," he explained, "we face a situation": Nixon, "Remarks to Midwestern News Media Executives," 807.

150 "I think of what happened to Greece and Rome": Nixon, "Remarks to Midwestern News Media Executives," 812.

151 Nixon called Connally to ask him to gather his thoughts on the currency situation and to approach it "with an open mind": Nichter, *Richard Nixon and Europe*, 50.

151 Peterson asked for permission: Nichter, *Richard Nixon and Europe*, 53.

151 The next day, Nixon met with Peterson and Connally together: Nichter, *Richard Nixon and Europe*, 53. Account of their conversation as described in this paragraph, including the quote.

152 On July 28, 1971, the *Wall Street Journal* reported of rumors in Paris: "Price of Gold Soars on Global Markets Following U.S. Report," *Wall Street Journal*, July 28, 1971.

152 "There are growing expectations in Europe": Clyde H. Farnsworth, "Gold Price Rises to a 2⊠Year High," *New York Times*, July 28, 1971.

152 the media kept accentuating negative developments in the United States: Clyde H. Farnsworth, "Health of Dollar Stirs New Worry," *New York Times*, July 30, 1971.

152 It was the first merchandise deficit since 1893: Peterson, "The United States in the Changing World Economy," 10; Matusow, *Nixon's Economy*, 145.

152 a four-hour meeting between Nixon and Connally, Shultz, and Haldeman in the Oval Office changed everything: Between August 2 and August 13 there were a variety of meetings. I relied heavily on Nichter, *Richard Nixon and Europe*; Douglas Brinkley and Luke A.

Nichter, eds., *The Nixon Tapes: 1971–1972* (Boston: Mariner Books, Houghton Mifflin Harcourt, 2014); Ritter, "Closing the Gold Window"; Silber, *Volcker*; Paul Volcker's Papers at the New York Federal Reserve Archive; and H. R. Haldeman, H. R. Haldeman Diaries Collection, Richard Nixon Presidential Library and Museum, Yorba Linda, CA [hereafter "Haldeman Diaries Collection RMN Library"].

152   This was essentially Volcker's latest contingency plan, updated as of July 27: Silber, *Volcker*, 83.

154   "I knew that by nature, he [Connally] always favored the 'big play'": Nixon, *RN*, 518.

154   "I am not sure this program will work": Nixon, *RN*, 518.

154   "This ought to show people you are aware of the problems": Ohlmacher, "The Dissolution of the Bretton Woods System Evidence from the Nixon Tapes, August–December 1971" (honor's thesis, University of Delaware, 2009), 10.

154   "[You] take a position before you are forced to": Ohlmacher, "The Dissolution of the Bretton Woods System Evidence from The Nixon Tapes," 10.

154   "It'll be as big a coup as your China thing": Ritter, "Closing the Gold Window," 238–39.

156   "This becomes a rather momentous decision": Haldeman, *Haldeman Diaries,* 335–36 (August 2, 1971).

157   On Wednesday, August 4, Nixon, Connally, and Shultz met again in the Oval Office: Ritter, "Closing the Gold Window," 243–45.

157   Nixon said that all deliberations on this economic package: "Editorial Note," doc. 164 in *Foreign Relations of the United States, 1969–1976*, vol. 3, *Foreign Economic Policy; International Monetary Policy, 1969–1972* (Washington, DC: U.S. Government Printing Office, 2001).

157   Wholesale prices were increasing at an annual rate of 8 percent: Details on wage settlement from Silk, *Nixonomics*, 71.

158   "If the Fund [fails] to meet its responsibility": "Action Now to Strengthen the U.S. Dollar: Report of the Subcommittee on International Exchanges and Payments of the Joint Economic Committee, Congress of the United States," August 1971 (Washington, DC: U.S. Government Printing Office, 1971), 13.

158   "If the market needed to be convinced that the dollar would be devalued": Allan H. Meltzer, *A History of the Federal Reserve*, Vol. 2, Book 2, 1970–1986 (Chicago: University of Chicago Press, 2009), 751.

158   Nixon warmed to the Texan's enthusiasm: Nichter, *Richard Nixon and Europe*, 56.

158 Five days later, Nixon met alone with Shultz: Ritter, "Closing the Gold Window," 246. Account of this conversation.

160 "The current flows of funds to other [financial] centers": Paul McCracken, "Memorandum for the President," August 9, 1971, FRASER, https://fraser.stlouisfed.org/archival/1173#3382.

160 demanding gold for dollars reached a new pitch: Volcker and Gyohten, *Changing Fortunes*, 76.

161 "Markets suggest that we are now at the point of witnessing that most dramatic of all confrontations": C. Gordon Tether, "$—When Something Has Got to Give," *Financial Times*, August 10, 1971.

161 "Thus on the one hand we have the U.S. authorities continuing to insist": C. Gordon Tether, "$—When Something Has Got to Give."

161 "Should the dollar be devalued": Dan Dorfman, "Heard On the Street," *Wall Street Journal*, August 11, 1971.

161 During the week of August 9, $4 billion in speculative money fled the United States: Coombs, *The Arena of International Finance*, 217.

161 foreign central banks absorbed another $1 billion in short-term inflows of dollars: James, *International Monetary Cooperation Since Bretton Woods*, 218.

161 on the morning of August 12, speculation against the dollar was so strong: Robert D. Hormats, "Information Memorandum from Robert Hormats of the National Security Council Staff to the President's Assistant for National Security Affairs (Kissinger), August 13, 1971," doc. 167 in *Foreign Relations of the United States, 1969–1976*, vol. 3, *Foreign Economic Policy; International Monetary Policy, 1969–1972* (Washington, DC: U.S. Government Printing Office, 2001).

161 it nevertheless created an air of panic: See Coombs, *The Arena of International Finance*, 218; Nichter, *Richard Nixon and Europe*, 57–93.

162 "One thing that was clear to me": Volcker and Gyohten, *Changing Fortunes*, 76.

162 Connally arrived at the Old Executive Office Building at 5:30 p.m. and joined Nixon and Shultz in the president's hideaway office: Silber, *Volcker*, 84–85; Ritter, "Closing the Gold Window," 257; Brinkley and Nichter, eds., *The Nixon Tapes: 1971–1972*, 231–61. Description of meeting and dialogue exchanges through the end of this section.

162 "I think we ought to be primarily concerned about how effectively you convince the American people": Quoted in Brinkley and Nichter, *The Nixon Tapes*, 235 (August 12, 1971).

163 The program will be picked apart: Quoted in Ritter, "Closing the Gold Window," 261.

163 "The biggest thing will be the impact on the American people": Quoted in Brinkley and Nichter, *The Nixon Tapes*, 240.

164 "I think that it's like the China announcement": Brinkley and Nichter, *The Nixon Tapes*, 243.

164 "Now is the time for decision . . . And off we go": Brinkley and Nichter, *The Nixon Tapes*, 252.

165 he had the chance to establish his leadership in domestic and international issues: Brinkley and Nichter, *The Nixon Tapes*, 250.

165 "This is the biggest step in economic policy since the end of World War II": Brinkley and Nichter, *The Nixon Tapes*, 260.

165 "It will be a shot heard round the world": Brinkley and Nichter, *The Nixon Tapes*, 252.

165 "I'll be the first to admit I'm wrong": Brinkley and Nichter, *The Nixon Tapes*, 260.

165 "We'll cover the [whole gamut of policies] when we do it": Haldeman, *Haldeman Diaries*, 340 (August 13, 1971).

Chapter 12: Friday, August 13

Most of this chapter and the two that follow are based on diaries and written contemporaneous accounts of Nixon chief of staff H. R. Haldeman, White House speechwriter William Safire, and Federal Reserve chairman Arthur Burns, all of whom were participants in the meeting. In addition, I interviewed the following men, listed here with titles they held at the time of the Camp David meeting: Undersecretary of the Treasury Paul Volcker, Director of the Office of Management and Budget George Shultz, OMB associate directors Arnold Weber and Kenneth Dam, and Treasury Deputy Counsel Michael Bradfield, all of whom also attended the Camp David meeting. I further relied on written recollections of Chairman of the Council of Economic Advisers Paul McCracken, CEA member Herbert Stein, assistant to the president Peter Peterson, chief of staff H. R. Haldeman, and Director of the Domestic Council John Ehrlichman, all of whom also were in the meetings. I interviewed, as well, Treasury Assistant Secretary John Petty, NSC staff member Robert Hormats, and Treasury Counsel Alan Wolff, all three of whom "held the fort" back in Washington during the meeting and were present in the White House on Sunday for the press conferences and the president's speech. A number of secondary sources were also used, including biographies and memoirs of Nixon, Connally, Burns, Volcker, and Kissinger and others noted in the footnotes, plus an article in *Life* magazine written shortly after the event and based on interviews of the weekend by White House correspondent Hugh Sidey. Not all firsthand and secondary accounts of the three days match up precisely, but I have done my best to reconcile them, omit the outliers, and otherwise interpret what took place. When it comes to dialogue, no tapes or official transcriptions of the meeting exist, but Safire took shorthand and transcribed sections of the dialogue in his book *Before the Fall*. I have included his verbatim dialogue only when it seemed accurate in the context of all the other available information.

166 "He was concerned about the fuzzy thinking when you get down to the nut cutting": Haldeman, diary entry, August 13, 1971, Haldeman Diaries Collection RMN Library, 1.

168 "The imposition of the [wage and price] freeze was a jump off the diving board": Stein, *Presidential Economics*, 177.

168 "There was no consensus within our government": Shultz and Dam, *Economic Policy Beyond the Headlines*, 119.

171 Many years later, Shultz recalled that he was satisfied: Characterization of Shultz's recollections from author's interview of George Shultz, August 16, 2017.

172 "In every economic speech I had ever worked on with him": Safire, *Before the Fall*, 509.

172 "Circumstances change": Safire, *Before the Fall*, 509.

173 His feet resting on a small footstool: Hugh Sidey, "The Economic Bombshell," *Life*, August 27, 1971, 20.

173 It is likely that Volcker had in his hand a Treasury report prepared that weekend: "U.S. Gold Stock and World Monetary Gold Holdings," August 15, 1971, FRC Box 7, Treasury, Records of Secretary of the Treasury George P. Shultz, 1971–74, National Archives, Washington, DC.

174 leading that year to the first trade deficit since 1893: Peterson, "The United States in the Changing World Economy," 10.

174 Volcker told the group a harrowing story: Coombs, *The Arena of International Finance*, 218; Meltzer, *A History of the Federal Reserve*, 2:754; Brandon, *The Retreat of American Power*, 225.

175 "He referred to China": Haldeman, diary entry, August 13, 1971, Haldeman Diaries Collection RMN Library, 2.

176 "He said it would not be relevant to know how we got here": Haldeman, diary entry, August 13, 1971, Haldeman Diaries Collection RMN Library, 2.

177 "Such a program will leave a clear impression that this [plan] has been analyzed in depth": Haldeman, diary entry, August 13, 1971, Haldeman Diaries Collection RMN Library, 3.

177 "It would be an act of great awareness, great statesmanship": Haldeman, diary entry, August 13, 1971, Haldeman Diaries Collection RMN Library, 2.

178 "The men around [Nixon] were to be the tacticians in a campaign": Sidey, "The Economic Bombshell," 20.

179 "No," he said. "That smacks wrong from the point of view of international leadership": Safire, *Before the Fall*, 512.

180 "No, let's finish with the border tax first": This exchange from Safire, *Before the Fall*, 513.

181 "Arthur, your view, as I understand it": Safire, *Before the Fall*, 513.

181 "We release forces we don't need to release": Haldeman, diary entry, August 13, 1971, Haldeman Diaries Collection RMN Library, 2.

181 He suggested that Paul Volcker be sent on a mission: Safire, *Before the Fall*, 513.

181 "I have never seen so many intelligent experts who disagree 180 degrees": Safire, *Before the Fall*, 514.

182 "What's our immediate problem?": Safire, *Before the Fall*, 514.

182 "I hate to do this, to close the window": The dialogue in this paragraph from Safire, *Before the Fall*, 514.

183 "McCracken was taking so many notes that he ran out of tablet paper": Sidey, "The Economic Bombshell."

184 A ten-minute speech would do it—"crisp, strong, confident": Haldeman, diary entry, August 13, 1971, Haldeman Diaries Collection RMN Library, 3.

185 the general reaction to closing the gold window would be negative: Description in this paragraph from Safire, *Before the Fall*, 514–15.

185 Connally intervened: Safire, *Before the Fall*, 515.

185 "they'll nibble us to death": Haldeman, diary entry, August 13, 1971, Haldeman Diaries Collection RMN Library, 4.

186 "Then [the president] issued an immortal quote": Haldeman, diary entry, August 13, 1971, Haldeman Diaries Collection RMN Library, 4.

188 Nixon returned to the speech he would give: The following paragraph consolidates and paraphrases notes from Safire, *Before the Fall*, 517; and Haldeman, diary entry, August 13, 1971, Haldeman Diaries Collection RMN Library, 12–13.

189 "When you background on this," said Nixon, "put 75 percent of your effort on TV": Safire, *Before the Fall*, 517.

189 He expressed concern: Burns, *Inside the Nixon Administration*, 52 (August 22, 1971).

190 He wrote in his diary: Burns, *Inside the Nixon Administration*, 53 (August 22, 1971).

190 "With that pledge, Burns gave Nixon what he wanted from Camp David above all": Matusow, *Nixon's Economy*, 153.

191 As dinner proceeded at Laurel: Haldeman, diary entry, August 13, 1971, Haldeman Diaries Collection RMN Library, 6.

191 "Exactly how?": Safire, *Before the Fall*, 518.

191 Shultz replied that it was about $23 billion: Summary of conversation between Volcker and Shultz from Safire, *Before the Fall*, 518.

192 "I began to thoroughly enjoy flaying the Secretary of the Treasury": Safire, *Before the Fall*, 519.

192 telling him: Ensuing dialogue in this paragraph from Safire, *Before the Fall*, 519.

192 Volcker made a phone call to Michael Bradfield: Author's interview of Michael Bradfield, July 12, 2017.

192 "Nobody had any intention of meeting with him further tonight": Haldeman, diary entry, August 13, 1971, Haldeman Diaries Collection RMN Library, 6.

Chapter 13: Saturday, August 14

194 with projections that temperatures would get to the high 70s or low 80s: "The Weather," *Frederick News Post*, August 14, 1971.

194 He filled both sides of three sheets with notes: Description of writing from Safire, *Before the Fall*, 519.

194 At 4:30 a.m., he called Haldeman: Account of phone conversation from Haldeman, diary entry, August 14, 1971, Haldeman Diaries Collection RMN Library, 1.

195   The amused president asked the sailor to take the tapes to Witch Hazel: Account of the pool interaction from Safire, *Before the Fall*, 518.

195   "He had a great line for Safire": Haldeman, diary entry, August 14, 1971, Haldeman Diaries Collection RMN Library, 1.

196   "[The president] said the major problem this morning was the gold float": Haldeman, diary entry, August 14, 1971, Haldeman Diaries Collection RMN Library, 1.

196   Nixon then came up with a phrase that we need to "defend the dollar against the international speculators": Haldeman, diary entry, August 14, 1971, Haldeman Diaries Collection RMN Library, 1.

196   He relayed the following directions to Haldeman: Haldeman, diary entry, August 14, 1971, Haldeman Diaries Collection RMN Library, 2.

197   "the exchange rate discrimination against the U.S.": Haldeman, diary entry, August 14, 1971, Haldeman Diaries Collection RMN Library, 2.

197   "Nixon came out of the TV studio sweating profusely, knowing he had 'lost'": Safire, *Before the Fall*, 3.

198   Nixon was the first president to build a large team of speechwriters: Author's interviews of Lee Huebner, November 15, 2017, and February 1, 2018; David Gergen, *Eyewitness to Power: The Essence of Leadership, Nixon to Clinton* (New York: Simon and Schuster, 2000), 53.

198   In March 1971, Safire was in such disfavor with Nixon: Safire, *Before the Fall*, 350.

198   (Safire himself did not discover the wiretapping until mid-1973): Safire, *Before the Fall*, 345.

199   "Once he got hold of the text": Author's interviews of Lee Huebner, November 15, 2017, and February 1, 2018.

199   a coach to his speechwriters: Author's interviews of Lee Huebner, November 15, 2017, and February 1, 2018; David Gergen, An Oral History Interview with David Gergen, interview by Timothy J. Naftali, audio, August 5, 2009, RMN Library.

199   he often seemed more concerned with the presentation of policy: Gergen, *Eyewitness to Power*, 55.

200   "His most important achievements": Reeves, *President Nixon* [e-book], "Introduction."

200   Nixon gave thirty-seven speeches from the Oval Office: Figure cited in Richard M. Nixon, "Address to the Nation Announcing Decision to Resign the Office of the President of the United States, August 8, 1974," American Presidency Project, https://www.presidency.ucsb.edu/documents

/address-the-nation-announcing-decision-resign-the-office-president-the
-united-states.

200   he went on TV to defend himself: Lee Huebner, "The Checkers Speech
After 60 Years," *The Atlantic*, September 22, 2012.

201   "to be hated and beaten": Safire, *Before the Fall*, 343.

201   the president told his staff to bar the reporter from entering the White
House: Jon Marshall, "Nixon Is Gone, but His Media Strategy Lives
On," *The Atlantic*, August 4, 2014.

201   "led to speculation that the administration's [domestic and foreign
economic] problems were undergoing review": Don Oberdorfer,
"Nixon's Economic Advisers Called to Weekend Sessions," *Washington
Post*, August 14, 1971.

202   "chaotic trading on world currency exchanges": H. Erich Heinemann,
"Chaotic Trading Weakens the Dollar," *New York Times*, August 14, 1971.

202   "on both sides of the Atlantic, there were rumors in the financial
community": Heinemann, "Chaotic Trading Weakens the Dollar."

202   Camp David proved an ideal setting: In this section, I was greatly
helped by three books on Camp David: W. Dale Nelson, *The President
Is at Camp David* (Syracuse, NY: Syracuse University Press, 1995);
Michael Giorgione, *Inside Camp David: The Private World of the
Presidential Retreat* (New York: Little, Brown and Company, 2017);
Jack Behrens, *Camp David Presidents: Their Families and the World*
(Bloomington, IN: AuthorHouse, 2014); and by H. R. Haldeman, Box
175, Alpha Subject Files, White House Special Files, Staff Member
Office Files, RMN Library.

202   "The Camp David establishment was arranged to give the participants
the sense of their unique value": Stein, *Presidential Economics*, 176.

203   "I never prepared for an important speech or press conference or made
a major decision in the oval office": Richard M. Nixon, *In the Arena:
A Memoir of Victory, Defeat, and Renewal* (New York: Simon and
Schuster, 1990), 162.

203   "In the mountains, Nixon was forever plotting": Cited in Giorgione,
Inside Camp David, 157.

204   When Nixon came by helicopter, as he usually did, a decoy chopper
often accompanied Marine One: Author's interviews of Camp David
guards Charles Nolan, January 19, 2018, and Dennis Morris, January
19, 2018.

204   "If by accident or intention, you turn in and proceed a few yards down
the path, everything changes": Michael Giorgione, "Inside Camp David,"

article excerpt, *The National*, n.d., http://www.amtrakthenational.com
/inside-camp-david.

204 In 1971, the security team at the camp was comprised of approximately
one hundred specially selected and trained Vietnam marine veterans:
Details in this paragraph from author's interviews of Charles Nolan,
January 19, 2018, and Dennis Morris, January 19, 2018.

205 After breakfast, three working groups met in three separate rooms in
Laurel Lodge: I was guided in the following section by handwritten
notes of John Ehrlichman found in John D. Ehrlichman, Box 6, Notes
of Meetings with President, 1969–1973, Staff Member Office Files,
White House Special Files, RMN Library; plus notes from Haldeman
and Safire in same file.

205 "We were winging it": Safire, *Before the Fall*, 522.

205 "But that was the only way it could have been done": Safire, *Before the
Fall*, 522.

207 "On the fifteenth day of the eighth month": Safire, *Before the Fall*, 522.

208 He mentioned that he would announce the imposition of a surcharge
on imports, but he did so in an elliptical way: Kissinger, *White House
Years*, 954.

208 "The fact was that a decision of major foreign policy importance had
been taken": Kissinger, *White House Years*, 954.

208 Haig in turn told Hormats, Kissinger's chief economic staff person,
to be on alert: Author's interviews of Robert Hormats, July 12 and
December 11, 2017.

209 "As I worked with Bill Safire that weekend," Nixon recalled in his
memoirs, "I wondered how the headlines would read": Nixon, *RN*, 520.

210 "The implications of slamming a lid on the American economy was [*sic*]
staggering": Safire, *Before the Fall*, 522.

210 "Volcker was undergoing an especially searing experience": Safire, *Before
the Fall*, 518.

210 Later that evening, Safire showed the entire speech to George Shultz
and also to Herb Stein: Safire, *Before the Fall*, 524.

211 "We walked in and the living room was empty": Haldeman, *Haldeman
Diaries*, 346 (August 14, 1971).

211 "The Japs, Russians, Chinese and Germans still have a sense of destiny
and pride": This quote and the one immediately following from
Haldeman, *Haldeman Diaries*, 346.

212 "He's a President now": Safire, *Before the Fall*, 524

212 "All in all, there was very little room for any doubt—taking the

president's words as he moved from one subject to the next—that he was governed mainly, if not entirely, by a political motive": Burns, *Inside the Nixon Administration*, 53 (August 22, 1971).

Chapter 14: Sunday, August 15

213 Afterward, he met with the entire team for about ninety minutes: Author's interview of Michael Bradfield, July 12, 2017.

213 make sure that both Connally and Burns got a lot of public credit: Haldeman, diary entry, August 15, 1971, Haldeman Diaries Collection RMN Library, 1.

213 he took time to admonish the Filipino orderly: Author's interview of Michael Bradfield, July 12, 2017.

215 he changed "tough" competition to "strong" competition: Safire, *Before the Fall*, 525.

215 When Safire heard it, he felt vaguely uncomfortable: This paragraph drawn from Safire, *Before the Fall*, 525.

215 "You know when all this was cooked up?": Safire, *Before the Fall*, 527.

215 "Nixon hated to do it": Safire, *Before the Fall*, 527.

216 he needed help from Volcker or Dam, then passed the cables through the State Department: Account from author's interviews of Robert Hormats, July 12 and December 11, 2017.

216 The cable from the president to West German chancellor Willy Brandt: Richard Nixon, "Telegram From the Department of State to the Embassy in Germany, August 16, 1971," doc. 169 in *Foreign Relations of the United States, 1969–1976*, vol. 3, *Foreign Economic Policy; International Monetary Policy, 1969–1972* (Washington, DC: U.S. Government Printing Office, 2001).

217 Another person who was scrambling that afternoon was John Petty: Author's interviews of John Petty, October 11 and November 9, 2017.

217 It was a historic and dramatic communication: Papers of Paul A. Volcker, Box 108477, New York Federal Reserve Archive.

217 At about 5:00 p.m., Nixon called John Mitchell: Timeline and quotes from Nichter, *Richard Nixon and Europe*, 66.

218 "The President has for a long time been considering": John B. Connally, George P. Shultz, and Paul W. McCracken, "Press Conference of Hon. John M. [*sic*] Connally, Secretary of the Treasury; Hon. George P. Shultz, Director, Office of Management and Budget; and Hon. Paul W. McCracken, Chairman, Council of Economic Advisers, The East

Room, August 15, 1971," August 15, 1971, FRC Box 7, Records of
Secretary of the Treasury George P. Shultz, 1971–1974, Record Group
56, National Archives at College Park, College Park, MD.

218   Questions from the media immediately went to the heart of the matter:
See Connally, Shultz, and McCracken, "Press Conference of Hon.
John M. [*sic*] Connally, Secretary of the Treasury; Hon. George P.
Shultz, Director, Office of Management and Budget; and Hon. Paul
W. McCracken, Chairman, Council of Economic Advisers, The East
Room, August 15, 1971."

219   On Sunday morning she received a call from McCracken's assistant:
Details on von Neumann Whitman from author's interview with
Marina von Neumann Whitman, December 2, 2017, and Marina von
Neumann Whitman, *The Martian's Daughter: A Memoir* (Ann Arbor,
MI: University of Michigan Press, 2012), 125.

219   That Sunday afternoon, Herb Stein called Alan Greenspan: Recounting
of the conversation between Herb Stein and Alan Greenspan from
author's interview of Alan Greenspan, November 15, 2017.

220   Secretary Rogers called Prime Minister Eisaku Satō of Japan to warn
him: Matusow, *Nixon's Economy*, 168.

220   Paul Volcker called Yasuke Kashiwagi: Author's interview of Paul
Volcker, July 26, 2017.

220   "We called this the Nixon Shock": Volcker and Gyohten, *Changing
Fortunes*, 92.

221   Dressed in a gray suit: Description of speech based on the video of
the speech ("The Challenge of Peace: Nixon's New Economic Policy")
available at https://youtu.be/ye4uRvkAPhA.

221   His copy of the speech was structured in outline form: Details of speech
from President's Personal File, Box 68, President's Speech File, RMN
Library.

221   "Address to the Nation Outlining a New Economic Policy: The
Challenge of Peace": Richard M. Nixon, "Address to the Nation
Outlining a New Economic Policy: The Challenge of Peace," doc. 264
in *Public Papers of the Presidents of the United States: Richard Nixon,
1971*.

223   "Without visible regret, certainly without apology": Safire, *Before the
Fall*, 527.

224   Safire came up with "Job Development": Safire, *Before the Fall*, 522.

228   "[i]mposing a freeze on wages and prices dramatized the president's anti-
inflation policy": Silk, *Nixonomics*, 64.

231 "Nixon was at his best when he was on the attack": Huebner's observations from author's interviews with Lee Huebner, November 15, 2017, and February 1, 2018.

238 Administration estimates showed that 46,200,000 Americans tuned in: "Memorandum for the President from Herbert C. Klein, August 16, 1971," Box 68, Presidential Speech File, President's Personal Files, RMN Library.

239 Petty, embarrassingly contrite, asked Schweitzer if he wanted to discuss the speech: Description of Connally, Volcker, Petty, and Schweitzer watching the speech from author's interviews of John Petty, October 11 and November 9, 2017.

239 Years later, Volcker reflected on his own reaction to the speech: Quotes and description in this paragraph from Volcker and Gyohten, *Changing Fortunes*, 79–80.

239 the three major networks provided short commentary for about ten to fifteen minutes each: "President Nixon's Speech on the Economy" aka "Nixon Shock Speech," with NBC commentary; ABC commentary featuring Paul McCracken; CBS commentary (8/15/1971), WHCA-4582, White House Communications Agency Videotape Collection, RMN Library.

241 Sometime Sunday evening, Paul Volcker received a note from Haldeman's office: Papers of Paul A. Volcker, Box 108477, New York Federal Reserve Archive.

241 Unfortunately for Volcker, his six-foot, seven-inch frame was too big for the bed: "The New Activist in Central Banking," *BusinessWeek*, no. 2193, September 11, 1971.

Chapter 15: The Aftermath

245 On August 16 the stock market jumped 32 points: Matusow, *Nixon's Economy*, 156.

245 "encountered enthusiasm bordering on euphoria": Matusow, *Nixon's Economy*, 156.

245 "I've never seen anything this unanimous unless maybe it was [the reaction to] Pearl Harbor": Matusow, *Nixon's Economy*, 156.

245 "On every specific action taken by the President, a majority of the public approved": Papers of John B. Connally, Folder: "JBC Memorandum from the House," Lyndon Baines Johnson Presidential Library, Austin, TX.

245 "We unhesitatingly applaud the boldness with which the President

has moved": "Call to Economic Revival," *New York Times*, August 16, 1971.

246 "mending the nation's pocketbook could pay off at the polls as Peking never would": "Nixon's Grand Design for Recovery," *Time*, August 30, 1971.

246 "It's a historic initiative": "Exploring the New Economic World," *Time*, August 30, 1971.

246 "Finally the world has been forced to look the problem in the face": "The Dollar: A Power Play Unfolds," *Time*, August 30, 1971.

246 "The Democrats have been embarrassed by this President": "Nixon's Grand Design for Recovery."

246 "the U.S. [government] will be an unseen but very real presence at the bargaining tables": "The Drastic Plan to Save the Dollar," *BusinessWeek*, August 21, 1971.

246 "The big question will be whether the world will accept what emerges": "Problems of Attitude," *Wall Street Journal*, August 17, 1971.

246 "The hidden danger in the latest world monetary crisis": "The Dollar: A Power Play Unfolds."

246 "Sooner rather than later, and the sooner the better": Matusow, *Nixon's Economy*, 157.

247 "Robin Hood in reverse": Matusow, *Nixon's Economy*, 157.

247 Leonard Woodcock, president of the United Auto Workers, said he was ready to declare war: Damon Stetson, "Unions Reject No-Strike Appeal," *New York Times*, August 19, 1971.

247 When Nixon's announcement was made: For account of officials and heads of government cutting short their vacations, see "The Dollar: A Power Play Unfolds."

248 This frenetic buying: Matusow, *Nixon's Economy*, 168.

248 no amount of purchasing dollars could prevent the greenback from sinking against the yen: "Squeeze on Japan," *New York Times*, August 20, 1971.

249 "The Japanese were too naïve in believing": For quote and description in this paragraph about Japan's belief that the United States would not sever the dollar–gold link, see Volcker and Gyohten, *Changing Fortunes*, 91–95.

250 "The dollar has collapsed as a leading currency": From a summary of press reactions compiled by the German embassy, "German Press Review, August 18, 1971," RMN Library.

250 "Nixon's program . . . documents a relapse of the world's strongest economic power": "German Press Review, August 18, 1971."

250 "The immediate significance of the new program was its effect abroad": Kissinger, *White House Years*, 955.

250 no one knew quite what to make of him: Author's interview of Valéry Giscard d'Estaing, November 21, 2017.

250 "stunned and flabbergasted and appalled by his crafty methods": Brandon, *The Retreat of American Power*, 240.

250 "They were used to following rules of quiet dignity in their negotiations": Brandon, *The Retreat of American Power*, 240.

251 At the White House, Monday, August 16: Haldeman, diary entry, August 16, 1971, Haldeman Diaries Collection RMN Library.

251 a memorandum reporting on the stock market: Paul McCracken, "Memorandum for the President, August 16, 1971," RMN Library.

252 "There is a saying that there is nothing constant except change": John B. Connally, "Statement by Secretary of the Treasury John B. Connally at the Opening of a News Conference, August 16, 1971," Papers of John B. Connally, 49-191C, Press Conference—Major Economic Programs, Monday, 8/16/71, Lyndon Baines Johnson Presidential Library, Austin, TX.

252 Having flown all night, Paul Volcker landed in London: Ensuing description of Volcker's activities in London from author's interview of Paul Volcker, July 26, 2017; Volcker and Gyohten, *Changing Fortunes*, 81; Silber, *Volcker*, 92–93; "Memorandum of Conversation, Subject: President Nixon's New Economic Program, August 16, 1971," doc. 170 in *Foreign Relations of the United States, 1969–1976*, vol. 3, *Foreign Economic Policy; International Monetary Policy, 1969–1972* (Washington, DC: U.S. Government Printing Office, 2001).

255 "France was concerned that everything that the U.S. had achieved": "Memorandum of Conversation, Subject: President Nixon's New Economic Program, August 17, 1971," doc. 171 in *Foreign Relations of the United States, 1969–1976*, vol. 3, *Foreign Economic Policy; International Monetary Policy, 1969–1972* (Washington, DC: U.S. Government Printing Office, 2001).

255 "Mr. Volcker replied that we equally recognized the danger": "Memorandum of Conversation, Subject: President Nixon's New Economic Program, August 17, 1971."

256 He found Nixon in a state of elation: Author's interview of Henry Kissinger, July 10, 2018; Kissinger, *White House Years*, 955.

256 "We stirred them up a bit": Quoted in Dallek, *Nixon and Kissinger*, 318.

256 "Mr. President, without you the country would be dead": Quoted in Dallek, *Nixon and Kissinger*, 318.

256 Kissinger sensed he would need to be involved in the follow-up to the Camp David decisions: Author's interview of Henry Kissinger, July 10, 2018.

256 he organized a conference call with two trusted and experienced advisors: Description of the phone call and its intent from author's interviews of Robert Hormats, July 12 and December 11, 2017, and Richard Cooper, June 7, 2018.

257 "We can all take credit for the program": "Nixon's Grand Design for Recovery."

257 "You know what we've done. Now what do we do?": Author's interview of C. Fred Bergsten, October 5, 2017.

257 Bergsten was unnerved by Connally's strident nationalist attitude: Author's interview of C. Fred Bergsten, October 5, 2017.

258 "We have fought two costly and grueling wars": Quotes in this paragraph from Richard M. Nixon, "Address to the Congress on Stabilization of the Economy, September 9, 1971," doc. 287 in *Public Papers of the Presidents of the United States: Richard Nixon, 1971*.

258 "Now, if we give up too soon": This quote and following from Nichter, *Richard Nixon and Europe*, 74.

259 "We have to have a strong America": Richard M. Nixon, "The President's News Conference of September 16, 1971," doc. 292 in *Public Papers of the Presidents of the United States: Richard Nixon, 1971*.

259 The gathering took place in Lancaster House: Description of the meeting and its events in the following section from Volcker and Gyohten, *Changing Fortunes*, 81–83; Matusow, *Nixon's Economy*, 169; Silk, *Nixonomics*, 100–102; Martin Mayer, *The Fate of the Dollar* (New York: Times Books, 1980), 192–94; Solomon, *The International Monetary System, 1945–1981*, 193, 198–99; author's interview of Paul Volcker, July 26, 2017; Paul A. Volcker, "Telegram from the Embassy in the United Kingdom to the Department of State, September 17, 1971," doc. 175 in *Foreign Relations of the United States, 1969–1976*, vol. 3, *Foreign Economic Policy; International Monetary Policy, 1969–1972* (Washington, DC: U.S. Government Printing Office, 2001); John B. Connally, "Secretary Connally's Statement at G-10 Meeting," September 15, 1971, Papers of John B. Connally, 49-191C: G-10 Meetings—London 9/15, 16, 17/71, Lyndon Baines Johnson Presidential Library, Austin, TX; plus various newspaper accounts referenced separately.

260 "The danger is that intransigence by the United States could spill over": This quote and the following from Clyde H. Farnsworth, "Economic

Ministers Are Facing a Rough Monetary Confrontation in the Group of 10 Meeting in London," *New York Times*, September 15, 1971.

260 His remarks were characterized as "uncompromising, hardnosed": "Text of Connally Talk to Rich Nations in London," *Washington Post*, September 26, 1971.

261 "We believe in it, we fostered it": Connally, "Secretary Connally's Statement at G-10 Meeting."

261 "But no nation should, over any period of time": Connally, "Secretary Connally's Statement at G-10 Meeting."

261 "We have not come here with any precise plans or details worked out": Connally, "Secretary Connally's Statement at G-10 Meeting."

262 he hadn't any intention of changing his position "by one iota": "Group of 10 Fails to Win Monetary Accord as U.S. Holds Out for Full Nixon Package," *Wall Street Journal*, September 17, 1971.

262 "the Finance Ministers failing even to agree on the order in which they are to approach the problem": William Keegan, "Group of Ten Talks End in Gridlock," *Financial Times*, September 17, 1971.

262 "[i]t will almost certainly force foreign retaliation": "Bargaining Tool or Bludgeon?" *New York Times*, September 17, 1971.

262 "We had a problem and we are sharing it with the world": John M. Lee, "$13 Billion Gain Sought to Spur Payment to U.S.," *New York Times*, September 16, 1971.

262 A cabinet meeting chaired by Nixon a week later: The section on this cabinet meeting is based on Haldeman, diary entry, September 24, 1971, Haldeman Diaries Collection RMN Library; with one exception, from Burns, *Inside the Nixon Administration*, 57 (October 7, 1971 entry).

263 Arthur Burns recalled that he was the lone objector: Burns, *Inside the Nixon Administration*, 57 (October 7, 1971).

263 "everyone wants to get into the U.S. market": Haldeman, diary entry, September 24, 1971, Haldeman Diaries Collection RMN Library.

263 "We can't have a trade war": Haldeman, diary entry, September 24, 1971, Haldeman Diaries Collection RMN Library.

263 "The President closed on the note": Haldeman, diary entry, September 24, 1971, Haldeman Diaries Collection RMN Library.

263 a number of congressional committees held hearings on the New Economic Policy: This section was informed by author's interviews of former congressional aides: Robert Cassidy, former staff member, Senate Finance Committee (October 18, 2017); Frank Cummings, former staff

member to Sen. Jacob Javits (December 6, 2017); Jane D'Arista, former staff, Joint Economic Committee (October 12, 2017); James Galbraith, former staff, Joint Economic Committee (November 27, 2017); Ken Guenther, former staff member to Sen. Jacob Javits (December 6, 2017); Jerry Jasinowski, former staff, Joint Economic Committee (October 5, 2017); Tom Korologos, former congressional staff for President Nixon (October 6, 2017); Karin Lissakers, former staff, Senate Subcommittee on Foreign Economic Policy (October 19, 2017); and Ken McLean, former staff member to Sen. William Proxmire (October 6, 2017).

265   "Today the committee opens one of the most important sets of hearings in its history": *The President's New Economic Program: Hearings Before the Joint Economic Committee*, 92nd Cong., Part 1 (1971), August 19, 1971, 1.

265   "The President has just brought about a drastic change of course": *The President's New Economic Program.*

265   "We have entered a complicated yet, nonetheless, decisive period": *The International Implications of the New Economic Policy: Hearings Before the Subcommittee on Foreign Economic Policy of the Committee on Foreign Affairs*, 92nd Cong. (1971), September 16, 1971, 1.

265   In the Senate Subcommittee on International Trade held on Monday, September 13: The following dialogue between Ribicoff and Volcker from *International Aspects of the President's New Economic Policies.*

268   Shultz invited Connally over to his house: The account of the Shultz-Connally-Volcker saga is based on Matusow, *Nixon's Economy*, 172; Volcker and Gyohten, *Changing Fortunes*, 82; and author's interviews of George Shultz (August 16, 2017) and Paul Volcker (July 26, 2017).

268   "In my own mind, the Shultz 'bombshell'": Volcker and Gyohten, *Changing Fortunes*, 82.

269   "Just about everybody is very angry at the United States": Edwin L. Dale Jr., "Everyone Is Angry at the U.S. as I.M.F. Talks Begin," *New York Times*, September 26, 1971.

269   "The language of diplomats runs to understatement and euphemism": Leonard Silk, "Dollar Devaluation," *New York Times*, September 29, 1971.

270   If these two requirements were met, Connally said, the United States would lift the surtax: See John B. Connally, "Statement by the Governor of the Fund and Bank for the United States," in *International Monetary Fund Summary Proceedings of the Twenty-Sixth Annual Meeting of the Board of Governors, September 27–October 1, 1971*, 215–22.

270 "Mr. Connally left no doubt that the United States is in deadly earnest":
Quoted in Mayer, *The Fate of the Dollar*, 195.

Chapter 16: The Finishing Line

271 In late August, Peterson told the president that Connally and Volcker
did not have the capacity for broad thinking: Haldeman, diary entry,
August 26, 1971, Haldeman Diaries Collection RMN Library, 1.

272 He crystalized all the differences in positions that existed within
the administration and posed issues that had yet to be resolved:
Peter G. Peterson, "Memorandum to the President, Negotiating the
New Economic Policy Abroad, September 23, 1971," Council on
International Economic Policy file, RMN Library.

273 Kissinger said the president was totally unaware of what was happening
abroad: Nichter, *Richard Nixon and Europe*, 76.

273 "Forces within the U.S. government which will wish to squeeze":
Robert D. Hormats, "Information Memorandum from Robert Hormats
of the National Security Council Staff to the President's Assistant for
National Security Affairs (Kissinger), September 28, 1971, Subject:
Progress in Developing USG Positions on International Aspects of
NEP," doc. 182 in *Foreign Relations of the United States, 1969–1976*,
vol. 3, *Foreign Economic Policy; International Monetary Policy, 1969–
1972* (Washington, DC: U.S. Government Printing Office, 2001).

274 Burns in particular warned Kissinger: Author's interviews of Robert
Hormats, July 12 and December 11, 2017.

274 surcharge was beginning to hit Western European exports hard:
Matusow, *Nixon's Economy*, 173.

274 "I came to the view that some shock had probably been needed":
Kissinger, *White House Years*, 957.

275 "I have to meet more regularly with Connally": Nichter, *Richard Nixon
and Europe*, 84.

275 "I'm coming as your friend": Nichter, *Richard Nixon and Europe*, 89.

275 "You've now smashed the system": Nichter, *Richard Nixon and Europe*, 89.

276 "cool it," quit the "saber rattling," and "stop the don't-give-a-damn":
Ritter, "Closing the Gold Window," 332.

276 Nixon was also receiving cables: Nichter, *Richard Nixon and Europe*, 89.

277 Hormats advised Kissinger to persuade Shultz: Robert D. Hormats,
"Information Memorandum from Robert Hormats of the National
Security Council Staff to the President's Assistant for National Security

Affairs (Kissinger), Subject: Talking Points for your Meeting with George Shultz, November 1, 1971," doc. 188 in *Foreign Relations of the United States, 1969–1976*, vol. 3, *Foreign Economic Policy; International Monetary Policy, 1969–1972* (Washington, DC: U.S. Government Printing Office, 2001).

277   Shultz shared this view: Ritter, "Closing the Gold Window," 342.

277   "We are uniting all those countries against us": Quoted in Nichter, *Richard Nixon and Europe*, 88.

277   Kissinger was particularly worried about Connally's repeated public statements: Ritter, "Closing the Gold Window," 331–32.

277   meetings in the Oval Office on November 22, 23, and 24: Matusow, *Nixon's Economy*, 174–75; Brandon, *The Retreat of American Power*, 236; Solomon, *The International Monetary System, 1945–1981*, 201; Ritter, "Closing the Gold Window," 364–65; "Editorial Note," doc. 203 in *Foreign Relations of the United States, 1969–1976*, vol. 3, *Foreign Economic Policy; International Monetary Policy, 1969–1972* (Washington, DC: U.S. Government Printing Office, 2001).

278   "Make a deal on monetary things": Quoted in Ritter, "Closing the Gold Window," 364.

278   "Say the deal on trade will come later": Quoted in Ritter, "Closing the Gold Window," 364.

278   "I discuss something with Pompidou": Ritter, "Closing the Gold Window," 375.

278   "He's got to have something to be credible": Ritter, "Closing the Gold Window," 375.

279   "Nixon Is Hopeful on Money Talks": Edwin L. Dale Jr., "Nixon Is Hopeful on Money Talks," *New York Times*, November 25, 1971.

279   The Rome meeting was held at the Palazzo Corsini: This section on the Rome meetings is based on author's interviews of Paul Volcker (July 26, 2017) and Valéry Giscard d'Estaing (November 21, 2017); Volcker and Gyohten, *Changing Fortunes*, 86–87; Solomon, *The International Monetary System, 1945–1981*, 201; Silber, *Volcker*, 99–100; Matusow, *Nixon's Economy*, 175; Reston, *The Lone Star*, 422–25; Ritter, "Closing the Gold Window," 387–408; Nichter, *Richard Nixon and Europe*, 91; Brandon, *The Retreat of American Power*, 239; and "Editorial Note," doc. 210 in *Foreign Relations of the United States, 1969–1976*, vol. 3, *Foreign Economic Policy; International Monetary Policy, 1969–1972* (Washington, DC: U.S. Government Printing Office, 2001).

279 "that President Nixon is now personally anxious to resolve": Paul Lewis, "U.S. Monetary Views More Flexible," *Financial Times*, November 29, 1971.

279 "Mr. Nixon, who cannot hope to present himself": This quote and the following from "Nixon's New Summitry," *Financial Times*, November 30, 1971.

281 Volcker then addressed the group: Description and account of Volcker's comments and Connally's are from Volcker and Gyohten, *Changing Fortunes*, 85–86.

281 "There was no answer, no discussion": Volcker and Gyohten, *Changing Fortunes*, 86.

281 He said that if the United States devalued against gold by 10 percent: Silber, *Volcker*, 100.

281 he could never agree to a U.S. devaluation of 10 percent: Reston, *The Lone Star*, 424.

282 Connally asked, "What would work?": Exchange between Barber and Connally from Reston, *The Lone Star*, 424.

282 "moment of the formal dethronement of the Almighty Dollar": Brandon, *The Retreat of American Power*, 239.

282 The Nixon-Pompidou summit: Sources for this section on the Azores include Volcker and Gyohten, *Changing Fortunes*, 87–88; Reston, *The Lone Star*, 425; Silber, *Volcker*, 101–2; Kissinger, *White House Years*, 959–69; Brandon, *The Retreat of American Power*, 241; Ritter, "Closing the Gold Window," 408–18; Haldeman, diary entry, December 13, 14, and 15, 1971, Haldeman Diaries Collection RMN Library; Tad Szulc, "Letter from the Azores," *The New Yorker*, January 1, 1972; various documents, Box 108477, Papers of Paul A. Volcker, Federal Reserve Bank of New York Archives; "Editorial Note," doc. 219 in *Foreign Relations of the United States, 1969–1976*, vol. 3, *Foreign Economic Policy; International Monetary Policy, 1969–1972* (Washington, DC: U.S. Government Printing Office, 2001); "'Framework for Monetary and Trade Settlement,' Paper Agreed by President Nixon and President Pompidou," doc. 220 in *Foreign Relations of the United States, 1969–1976*, vol. 3, *Foreign Economic Policy; International Monetary Policy, 1969–1972* (Washington, DC: U.S. Government Printing Office, 2001); author's interviews of Henry Kissinger (July 10, 2018), Valéry Giscard d'Estaing (November 21, 2017), Robert Hormats (July 12 and December 11, 2017), and Édouard Balladur (February 8, 2018).

283 "So it happened that a solution to the monetary crisis": Kissinger, *White House Years*, 960.

283 "Even in my most megalomaniac moments": Kissinger, *White House Years*, 960.

284 the United States should devalue by 6 percent: Brandon, *The Retreat of American Power*, 241.

284 "If given a truth serum": Kissinger, *White House Years*, 961.

284 At 4:20 a.m. Azores time: Haldeman, *Haldeman Diaries*, 384 (December 14, 1971).

284 the private agreement between the two presidents contained significant detail: "'Framework for Monetary and Trade Settlement,' Paper Agreed by President Nixon and President Pompidou."

285 "I think it is fair to say that the meeting between the two Presidents": "Remarks of the President, Secretary of the Treasury John B. Connally, and Secretary of State William P. Rogers on Arrival at Andrews Air Force Base, December 14, 1971," *Weekly Compilation of Presidential Documents* 7, no. 51 (December 20, 1971), 1665.

286 "The Group of Ten meeting which will occur this weekend": "Remarks of the President, Secretary of the Treasury John B. Connally, and Secretary of State William P. Rogers on Arrival at Andrews Air Force Base, December 14, 1971."

286 "The Azores agreement . . . is a momentous event": Hobart Rowen, "New Patterns for Money: President Abandons Historic Stand," *Washington Post*, December 15, 1971.

286 "The Azores communiqué is the economic counterpart of the Guam doctrine": "On Devaluation," *Washington Post*, December 15, 1971.

286 "It serves notice on the rest of the world": "On Devaluation."

286 In a morning meeting the next day: Haldeman, diary entry, December 15, 1971, Haldeman Diaries Collection RMN Library.

286 In his diary, Haldeman recorded: Haldeman, diary entry, December 15, 1971, Haldeman Diaries Collection RMN Library.

287 it was as close as the United States had to a genuine castle: Description of Connally and the Smithsonian from Silber, *Volcker*, 102; Volcker and Gyohten, *Changing Fortunes*, 88–89.

287 He received considerable support as well as a commitment for quick congressional action: Richard F. Janssen, "Nixon Agrees to Formally Devalue Dollar in Azores Announcement with Pompidou," *Wall Street Journal*, December 15, 1971.

287 CEOs cited the importance of ending the uncertainty: "The New Lineup: Devaluation of Dollar, Revaluation of Others Seen Aiding Economy," *Wall Street Journal*, December 15, 1971.

288 the dollar was down: Clyde H. Farnsworth, "Dollar Sags Sharply as Nations Move Closer to Monetary Pact," *New York Times*, December 16, 1971.

288 "We are prepared to push for a conclusion": Edwin L. Dale Jr., "Monetary Talks Make Slow Start: Group of 10 Finance Chiefs Meet— Connally Is Hopeful," *New York Times*, December 18, 1971.

290 "Everyone was too concerned with protecting his own position": Brandon, *The Retreat of American Power*, 241–42.

290 Connally recalled to writer Martin Mayer: Mayer, *The Fate of the Dollar*, 200.

290 "He cajoled, threatened and roughed up": This quote and following from Brandon, *The Retreat of American Power*, 241–42.

291 Connally couldn't get the Japanese to budge: The following account of the negotiation comes from Mayer, *The Fate of the Dollar*, 201; Reston, *The Lone Star*, 430–31; Volcker and Gyohten, *Changing Fortunes*, 96–97.

292 They had agreed to an exchange rate realignment: This section is based on Solomon, *The International Monetary System, 1945–1981*, 207–11; Meltzer, *A History of the Federal Reserve*, 2:776; James, *International Monetary Cooperation Since Bretton Woods*, 237; "Text of Group of Ten Communique," *Department of State Bulletin* 66, no. 1698 (January 10, 1972): 32–34.

292 the overall U.S. devaluation was a little shy of 8 percent: Solomon, *The International Monetary System, 1945–1981*, 207–11.

294 The president stood in front of the airplane the Wright brothers flew: Richard F. Janssen, "The New Dollar: Devaluation of 8.57% Likely to Create Jobs, Help Nixon Summitry," *Wall Street Journal*, December 20, 1971.

294 "It is my great privilege to announce": Richard M. Nixon, "Remarks Announcing a Monetary Agreement Following a Meeting of the Group of Ten, December 18, 1971," doc. 401 in *Public Papers of the Presidents of the United States: Richard Nixon, 1971*.

294 "The remark has been repeated with a scornful laugh": Volcker and Gyohten, *Changing Fortunes*, 90.

295 "When we compare this agreement to Bretton Woods": This quote and the following from Nixon, "Remarks Announcing a Monetary Agreement Following a Meeting of the Group of Ten, December 18, 1971."

295   "But we are pleased it's settled": This quote and the following from
      Hobart Rowen, "Monetary Agreement Reached: U.S. Devalues; Nixon
      Unveils Historic Pact," *Washington Post*, December 19, 1971.

296   "After so many months and so many uncertainties": Douglas W. Cray,
      "Pact Encourages Business Experts," *New York Times*, December 19,
      1971.

296   The chairman of IBM, T. Vincent Learson, praised the value: Cray,
      "Pact Encourages Business Experts."

296   praised the agreement as a "remarkably good job": William D. Smith, "U.S.
      Economists Hail Monetary Accord," *New York Times*, December 20,
      1971.

296   "Essentially, the agreement removes a gnawing uncertainty": Clyde H.
      Farnsworth, "European Hopes Buoyed," *New York Times*, December 19,
      1971.

296   There was a feeling that: Farnsworth, "European Hopes Buoyed."

296   "I hope it lasts three months": Volcker and Gyohten, *Changing Fortunes*, 90.

Chapter 17: The Long View

297   Immediately after the Smithsonian Agreement was concluded: Edwin
      L. Dale Jr., "Monetary Fund Approves World Currency Moves," *New
      York Times*, December 20, 1971.

297   "Nixon's unilateral decisions of August 15 had their desired effect":
      Kissinger, *White House Years*, 962.

297   U.S. stocks soared: Terry Robards, "Currency Accord Sends Stocks Up
      in Heavy Trading," *New York Times*, December 21, 1971.

297   "The remarkable events set in motion on August 15": Hobart Rowen,
      "Devaluation: Presidential Midas Touch," *Washington Post*, December
      20, 1971.

297   The *Wall Street Journal* cast doubt: "Very Like a Whale," *Wall Street
      Journal*, December 20, 1971.

298   "We should have had a 36 percent revaluation from the Japanese":
      Mayer, *The Fate of the Dollar*, 203.

298   "[The devaluation] was well short": Volcker and Gyohten, *Changing
      Fortunes*, 89.

298   "What was lacking was the sense of commitment": Volcker and
      Gyohten, *Changing Fortunes*, 90.

298   "Nixon, like most presidents": Volcker and Gyohten, *Changing Fortunes*,
      104.

299 "The U.S. must first of all keep the moral undertakings it has made": Adrian Dicks, "U.S. Must Abide by Its 'Moral Undertakings,' Says Pompidou," *Financial Times*, December 23, 1971.

299 "The new exchange rates for the dollar constitute a successful interim solution": "The Dollar Accord," *Washington Post*, December 21, 1971.

300 "The underlying problems that created the monetary crisis": Leonard Silk, "What Now for Dollar?" *New York Times*, December 22, 1971.

300 foreign governments that held some $60 billion: The $60 billion is Silk's number. Treasury said $40 billion.

300 the world now faced a system problem, not just a dollar problem: H. Erich Heinemann, "The Dollar as Kingpin Is Dead: Any New Monetary System Will Give It Smaller Role," *New York Times*, December 20, 1971.

301 The program bred the illusion: Matusow, *Nixon's Economy*, 198.

301 the president relentlessly pressured Arthur Burns: Abrams, "How Richard Nixon Pressured Arthur Burns," 177–88; Pierce, "The Political Economy of Arthur Burns," 485–96.

302 Fiscal and monetary stimuli were adding gasoline: Matusow, *Nixon's Economy*, 165.

302 "lash the economic system to a gallop": "Nixon's Go-Go Economic Policies for '72," *BusinessWeek*, January 29, 1972.

302 In the year following the Smithsonian Agreement: Silber, *Volcker*, 104–6.

303 The Western Europeans, in particular, were convinced: Solomon, *The International Monetary System, 1945–1981*, 217.

303 "I feel obligated to describe to you": "Letter from President Pompidou to President Nixon, February 4, 1972," doc. 223 in *Foreign Relations of the United States, 1969–1976*, vol. 3, *Foreign Economic Policy; International Monetary Policy, 1969–1972* (Washington, DC: U.S. Government Printing Office, 2001).

303 Nixon was unapologetic in his reply of Wednesday, February 16: "Letter from President Nixon to President Pompidou, February 16, 1972," doc. 224 in *Foreign Relations of the United States, 1969–1976*, vol. 3, *Foreign Economic Policy; International Monetary Policy, 1969–1972* (Washington, DC: U.S. Government Printing Office, 2001).

304 the British made the first formal break: The following discussion of the breakdown of the Smithsonian Agreement and the eventual legalization of floating exchange rates and reform of the "non-system" is based in good part on Volcker and Gyohten, *Changing Fortunes*, 101–28;

Solomon, *The International Monetary System, 1945–1981*, 216–34; and author's interview of George Shultz, August 16, 2017.

304 the White House tape recorder captured the following dialogue: Transcript of Richard Nixon tapes, Tape 741–002, June 23, 1972, Oval Office, 10:04–11:39, RMN Library. Conversation has been abridged for length.

305 his proposal made little progress: Volcker and Gyohten, *Changing Fortunes*, 122–23.

306 In five days, he traveled 31,000 miles: Solomon, *The International Monetary System, 1945–1981*, 230.

306 the Organization of Petroleum Exporting Countries (OPEC) raised oil prices: Solomon, *The International Monetary System, 1945–1981*, 258.

307 "In the process, for better or worse": Volcker and Gyohten, *Changing Fortunes*, 102.

308 "The right to float must be clear and unencumbered": Solomon, *The International Monetary System, 1945–1981*, 270.

308 "a decision to learn to live with a non-system": John Williamson, *The Failure of World Monetary Reform, 1971–1974* (New York: New York University Press, 1977), 73.

309 "It will be, in other words, a very permissive system": Leonard Silk, "Shaping World Economy: A Managed System of Floating Rates Appears to Be I.M.F.'s Accepted Plan," *New York Times*, January 7, 1976.

309 "All is well that ends": Quoted in Tom de Vries, "Jamaica, or the Non-Reform of the International Monetary System," *Foreign Affairs* 54, no. 3 (April 1976).

310 That agreement worked for a while: Author's interviews of former secretary of the treasury James Baker (November 28, 2017) and his undersecretary, David Mulford (December 5, 2017).

311 "Today we face a new Bretton Woods moment": Kristalina Georgieva, "A New Bretton Woods Moment," IMF Annual Meeting, October 15, 2020.

Chapter 18: The Weekend in Retrospect

313 "Bad medicine at a bad time": Shultz and Dam, *Economic Policy Beyond the Headlines*, 85.

314 "The August 15, 1971, decision to impose [wage and price controls]": This quote and the following from Nixon, *RN*, 521.

314   "But in the long run": Nixon, *RN*, 521.

314   In his 1971 study: Peterson, "The United States in the Changing World Economy."

317   "The Nixon Shock was a central cause of the Great Inflation": Roger Lowenstein, "The Nixon Shock," Bloomberg.com, August 4, 2011.

317   "the worst inflation in American history": Lewis E. Lehrman, "The Nixon Shock Heard 'Round the World," *Wall Street Journal*, August 15, 2011.

323   "The new system must be characterized by shared leadership": Peterson, "The United States in the Changing World Economy," v.

323   "It will be a system which fully recognizes": Peterson, "The United States in the Changing World Economy," v.

323   In think tanks and international fora around the world: Also see, for example: UN World Food Conference, Rome, 1974, which was preceded by multiple studies and meetings; the UN Conference on World Population, Bucharest, 1974, which was preceded by multiple international studies and meetings; the establishment of the International Energy Agency, Paris, 1974, focused on energy security and cooperation among major energy-consuming nations; and the establishment of the Independent Commission on International Development, 1977, chaired by former West German chancellor Willy Brandt.

BIBLIOGRAPHY

This bibliography is organized as follows: General Acknowledgments, Books, Articles and Book Chapters, Official Documents and Primary Materials, Author Interviews.

General Acknowledgments

Before noting the references I used, I would like to acknowledge that a specific set of sources was most useful to me. Most important were the interviews I conducted; see "Author Interviews," page 421. Other critical sources include: Robert Solomon, *The International Monetary System, 1945–1981*; Harold James, *International Monetary Cooperation Since Bretton Woods*; Michael D. Bordo and Barry Eichengreen, eds., *A Retrospective on the Bretton Woods System: Lessons for International Monetary Reform*; several books and numerous articles by C. Fred Bergsten (see "Articles and Book Chapters"); Allen Matusow, *Nixon's Economy: Booms, Busts, Dollars, and Votes*; William Safire, *Before the Fall: An Inside View of the Pre-Watergate White House*; Paul Volcker and Toyoo Gyohten, *Changing Fortunes: The World's Money and the Threat to American Leadership*, plus several other Volcker sources; Arthur Burns, *Inside the Nixon Administration: The Secret Diary of Arthur Burns, 1969–1974*; Richard Nixon, *RN: The Memoirs of Richard Nixon*; Henry Kissinger, *White House Years*; Henry Brandon, *The Retreat of American Power: The Inside Story of How Nixon and Kissinger Changed Foreign Policy for Years to Come*; and Luke A. Nichter, *Richard Nixon and Europe: The Reshaping of the Postwar Atlantic World*. I was assisted greatly by John Connally's papers at the LBJ Library; the Nixon, Peterson, Haldeman, and some McCracken papers at the Nixon Library; and Volcker's papers at the Federal Reserve Bank of New York. Two general reference works proved invaluable: the series *Foreign Relations of the United States, 1969–1976,* vol. 3, edited by the Department of State, and *Public Papers of the Presidents of the United States: Richard Nixon, 1969, 1970, and 1971,* compiled by the General Services Administration and available from the U.S. Government Printing Office. I benefited from a 1971 report by Peter Peterson called, "The United States in the Changing World Economy." Also essential were PhD dissertations by Joanne Gowa, "Explaining Large Scale Policy Change: Closing the Gold Window, 1971"; Christen Thomas Ritter, "Closing the Gold Window: Gold, Dollars, and the Making of Nixonian Foreign Economic Policy"; and Tom Forbord, "The Abandonment of Bretton Woods: The Political Economy of U.S. International Monetary Policy." Finally, I wish to acknowledge the help of the Beinecke Rare Book and Manuscript Library at Yale for several documents including various editions of *Fortune* magazine.

Books

Alden, Edward. *Failure to Adjust: How Americans Got Left Behind in the Global Economy.* Lanham, MD: Rowman and Littlefield, 2017.

Ambrose, Stephen E., and Douglas Brinkley. *Rise to Globalism: American Foreign Policy Since 1938.* 8th rev. ed. New York: Penguin Books, 1997.

Auletta, Ken. *Greed and Glory on Wall Street: The Fall of the House of Lehman.* New York: Random House, 1986.

Axilrod, Stephen H. *Inside the Fed: Monetary Policy and Its Management, Martin Through Greenspan to Bernanke.* Cambridge, MA: The MIT Press, 2011.

Behrens, Jack. *Camp David Presidents: Their Families and the World.* Bloomington, IN: AuthorHouse, 2014.

Bergsten, C. Fred. *The Dilemmas of the Dollar: The Economics and Politics of United States International Monetary Policy.* New York: New York University Press, 1975.

Bergsten, C. Fred, and Joseph Gagnon. *Currency Conflict and Trade Policy: A New Strategy for the United States.* Washington, DC: Peterson Institute for International Economics, 2017.

Bergsten, C. Fred, and Russell Aaron Green, eds. *International Monetary Cooperation: Lessons from the Plaza Accord After Thirty Years.* Washington, DC: Peterson Institute for International Economics, 2016.

Bergsten, C. Fred, and John Williamson, eds. *Dollar Overvaluation and the*

World Economy. Washington, DC: Peterson Institute for International Economics, 2003.

Bordo, Michael D., and Barry Eichengreen. *A Retrospective on the Bretton Woods System: Lessons for International Monetary Reform.* Chicago, IL: University of Chicago, 1993.

Brandon, Henry. *The Retreat of American Power: The Inside Story of How Nixon and Kissinger Changed Foreign Policy for Years to Come.* New York: Doubleday, 1973.

————. *Special Relationships: A Foreign Correspondent's Memoirs from Roosevelt to Reagan.* New York: Scribner's, 2015.

Brinkley, Douglas, and Luke A. Nichter, eds. *The Nixon Tapes: 1971–1972.* Boston: Mariner Books/Houghton Mifflin Harcourt, 2014.

Burns, Arthur F. *Inside the Nixon Administration: The Secret Diary of Arthur Burns, 1969–1974.* Edited by Robert H. Ferrell. Lawrence, KS: University of Kansas Press, 2010.

————. *Reflections of an Economic Policy Maker: Speeches, Congressional Statements: 1969–1978.* Washington, DC: American Enterprise Institute, 1978.

Caro, Robert A. *Means of Ascent: The Years of Lyndon Johnson.* New York: Alfred A. Knopf, 1990.

Connally, John B. *In History's Shadow: An American Odyssey.* New York: Hyperion, 1993.

Coombs, Charles A. *The Arena of International Finance.* New York: Wiley, 1976.

Cooper, Richard N. *The Economics of Interdependence: Economic Policy in the Atlantic Community.* Council on Foreign Relations Series. New York: Columbia University Press, 1968.

Dallek, Robert. *Nixon and Kissinger: Partners in Power.* New York: HarperCollins, 2007.

Destler, I. M. *American Trade Politics: System Under Stress.* Washington, DC: Institute for International Economics, 1986.

Drew, Elizabeth. *Richard M. Nixon.* New York: Times Books, 2007.

Ehrlichman, John. *Witness to Power: The Nixon Years.* New York: Simon and Schuster, 1982.

Farrell, John A. *Richard Nixon: The Life.* New York: Doubleday, 2017.

Ferguson, Niall. *Kissinger.* Vol. 1, *1923–1968: The Idealist.* New York: Penguin Press, 2015.

Fulbright, J. William. *The Arrogance of Power.* New York: Vintage Books, 1966.

Funabashi, Yōichi. *Managing the Dollar: From the Plaza to the Louvre.* 2nd ed. Washington, DC: Institute for International Economics, 1988.

Gagnon, Joseph E., and Marc Hinterschweiger. *Flexible Exchange Rates for a Stable World Economy.* Washington, DC: Peterson Institute for International Economics, 2011.

Gergen, David. *Eyewitness to Power: The Essence of Leadership, Nixon to Clinton.* New York: Simon and Schuster, 2000.

Giorgione, Michael. *Inside Camp David: The Private World of the Presidential Retreat.* New York: Little, Brown and Company, 2017.

Gordon, Kermit, ed. *Agenda for the Nation: Papers on Domestic and Foreign Policy Issues.* Washington, DC: Brookings Institution, 1968.

Greider, William. *Secrets of the Temple: How the Federal Reserve Runs the Country.* New York: Simon and Schuster, 1987.

Haldeman, H. R. *The Haldeman Diaries: Inside the Nixon White House.* New York: G. P. Putnam's, 1994.

Hargrove, Erwin C., and Samuel A. Morley, eds. *The President and the Council of Economic Advisers: Interviews with CEA Chairmen.* Boulder, CO: Westview Press, 1984.

Hughes, Ken. *Chasing Shadows: The Nixon Tapes, the Chennault Affair, and the Origins of Watergate.* Charlottesville: University of Virginia Press, 2014.

Ikenberry, G. John. *After Victory: Institutions, Strategic Restraint, and the Rebuilding of Order After Major Wars.* E-book. Princeton, NJ: Princeton University Press, 2001.

Irwin, Douglas A. *Clashing Over Commerce: A History of U.S. Trade Policy.* Chicago, IL: University of Chicago Press, 2017.

Isaacson, Walter. *Kissinger: A Biography.* New York: Simon and Schuster, 1992.

James, Harold. *International Monetary Cooperation Since Bretton Woods.* Washington, DC: International Monetary Fund and Oxford University Press, 1996.

James, Harold, and Juan Carlos Martinez Oliva, eds. *International Monetary Cooperation Across the Atlantic.* Frankfurt am Main: European Association for Banking and Financial History, 2007.

Jones, Sidney L. *Public and Private Economic Advisor: Paul W. McCracken.* Lanham, MD: University Press of America, 2000.

Kaufman, Henry. *On Money and Markets: A Wall Street Memoir.* New York: McGraw-Hill, 2000.

Kissinger, Henry. *White House Years.* Boston: Little, Brown and Company, 1979.

Levinson, Marc. *An Extraordinary Time: The End of the Postwar Boom and the Return of the Ordinary Economy*. New York: Basic Books, 2016.

Lord, Winston. *Kissinger on Kissinger: Reflections on Diplomacy, Grand Strategy, and Leadership*. E-book. New York: All Points Books, 2019.

Matusow, Allen J. *Nixon's Economy: Booms, Busts, Dollars, and Votes*. Lawrence, KS: University Press of Kansas, 1998.

Mayer, Martin. *The Fate of the Dollar*. First Signet Printing edition. New York: Times Books, 1980.

Meltzer, Allan H. *A History of the Federal Reserve*. Vol. 2, Book 2: *1970–1986*. 3 vols. Chicago: University of Chicago Press, 2009.

Nelson, W. Dale. *The President Is at Camp David*. Syracuse, NY: Syracuse University Press, 1995.

Nichter, Luke A. *Richard Nixon and Europe: The Reshaping of the Postwar Atlantic World*. Cambridge, UK, and New York: Cambridge University Press, 2017.

Nixon, Richard M. *In the Arena: A Memoir of Victory, Defeat, and Renewal*. New York: Simon and Schuster, 1990.

———. *RN: The Memoirs of Richard Nixon*. New York: Grosset and Dunlap, 1978.

Odell, John S. *U.S International Monetary Policy: Markets, Power, and Ideas as Sources of Change*. Princeton, NJ: Princeton University Press, 1982.

Okun, Arthur M. *The Political Economy of Prosperity*. New York: W. W. Norton and Company, 1970.

Peterson, Peter G. *The Education of an American Dreamer: How a Son of Greek Immigrants Learned His Way from a Nebraska Diner to Washington, Wall Street, and Beyond*. E-book. New York: Hachette Book Group, 2009.

Preeg, Ernest H. *The Trade Deficit, the Dollar, and the U.S. National Interest*. Indianapolis, IN: Hudson Institute, 2000.

Reeves, Richard. *President Nixon: Alone in the White House*. E-book. Simon and Schuster, 2001.

Reston, James, Jr. *The Lone Star: The Life of John Connally*. New York: Harper and Row, 1989.

Roosa, Robert V. *The Dollar and World Liquidity*. New York: Random House, 1967.

Rothkopf, David J. *Running the World: The Inside Story of the National Security Council and the Architects of American Power*. New York: Public Affairs, 2005.

Safire, William. *Before the Fall: An Inside View of the Pre-Watergate White House*. 1st ed. Garden City, NY: Doubleday, 1975.

Samuelson, Robert J. *The Great Inflation and Its Aftermath.* New York: Random House, 2010.

Scalapino, Robert. *American-Japanese Relations in a Changing Era.* Washington, DC: Georgetown University Center for Strategic and International Studies, 1972.

Sestanovich, Stephen. *Maximalist: America in the World from Truman to Obama.* New York: Vintage, 2014.

Shultz, George P. *Learning from Experience.* Stanford, CA: Hoover Institution Press, 2016.

———. *Turmoil and Triumph: My Years as Secretary of State.* E-book. New York: Scribner's, 1993.

Shultz, George P., and Kenneth Dam. *Economic Policy Beyond the Headlines.* Stanford, CA: Stanford Alumni Association, 1977.

Silber, William L. *Volcker: The Triumph of Persistence.* New York: Bloomsbury Press, 2012.

Silk, Leonard S. *Nixonomics: How the Dismal Science of Free Enterprise Became the Black Art of Controls.* New York: Praeger Publishers, 1972.

Solomon, Robert. *The International Monetary System, 1945–1981.* New York: Harper and Row, 1982.

Steil, Benn. *The Battle of Bretton Woods: John Maynard Keynes, Harry Dexter White, and the Making of a New World Order.* Princeton, NJ: Princeton University Press, 2013.

Stein, Herbert. *Presidential Economics: The Making of Economic Policy from Roosevelt to Reagan and Beyond.* New York: Simon and Schuster, 1984.

Strange, Susan, and Christopher Prout. *International Monetary Relations.* Vol. 2, *International Economic Relations of the Western World.* London: Oxford University Press, 1976.

Thomas, Evan. *Nixon: A Man Divided.* E-book. New York: Random House, 2015.

Van Overtveldt, Johan. *The Chicago School: How the University of Chicago Assembled the Thinkers Who Revolutionized Economics and Business.* Chicago: Agate B2, 2007.

Volcker, Paul, and Toyoo Gyohten. *Changing Fortunes: The World's Money and the Threat to American Leadership.* New York: Times Books, 1992.

Volcker, Paul A., and Christine Harper. *Keeping at It: The Quest for Sound Money and Good Government.* New York: PublicAffairs, 2018.

Wells, Wyatt C. *Economist in an Uncertain World: Arthur F. Burns and the Federal Reserve, 1970–78.* New York: Columbia University Press, 1994.

Whitman, Marina von Neumann. *The Martian's Daughter: A Memoir*. Ann Arbor, MI: The University of Michigan Press, 2012.

Williamson, John. *The Failure of World Monetary Reform, 1971–1974*. New York: New York University Press, 1977.

Articles and Book Chapters

Abrams, Burton A. "How Richard Nixon Pressured Arthur Burns: Evidence from the Nixon Tapes." *The Journal of Economic Perspectives* 20, no. 4 (2006): 177–88.

Abrams, Burton A., and James L. Butkiewicz. "The Political Economy of Wage and Price Controls: Evidence from the Nixon Tapes." *Public Choice* 170, no. 1 (January 2017): 63–78.

"The Administration Takes on the Fed." *BusinessWeek*, July 31, 1971.

Aldasoro, Iñaki, and Torsten Ehlers. "Global Liquidity: Changing Instrument and Currency Patterns." *BIS Quarterly Review* (September 2018). https://www.bis.org/publ/qtrpdf/r_qt1809b.htm.

"America Must Use Sanctions Cautiously." *The Economist*, May 17, 2018.

". . . And How to Negotiate It." *New York Times*, August 18, 1971, sec. Editorial.

Appelbaum, Binyamin, and Robert D. Hershey Jr. "Paul A. Volcker, Fed Chairman Who Waged War on Inflation, Is Dead at 92." *New York Times*, December 9, 2019, sec. Business.

Apple, R. W., Jr. "The 37th President; Richard Nixon, 81, Dies; A Master of Politics Undone by Watergate." *New York Times*, April 23, 1994.

"The Architect of Nixon's New Economics." *BusinessWeek*, March 20, 1971.

"Art Surrounds Monetary Talks: Many Masterpieces Add Glamour for the Group of 10." *New York Times*, December 2, 1971, sec. Business/Finance.

"Back and Forth on the Economy." *New York Times*, July 2, 1971, sec. Editorial.

"Bargaining Tool or Bludgeon?" *New York Times*, September 17, 1971, sec. Editorial.

Bergsten, C. Fred. "Crisis in U.S. Trade Policy." *Foreign Affairs (Pre-1986); New York* 49, no. 4 (July 1971): 619.

———. "Taking the Monetary Initiative." *Foreign Affairs* 46, no. 4 (July 1968): 713–32.

———. "Time for a Plaza II?" *International Monetary Cooperation: Lessons from the Plaza Accord After Thirty Years*. Edited by C. Fred Bergsten and Russell Aaron Green. Washington, DC: Peterson Institute for International Economics, 2016, 261–94

Bergsten, C. Fred, and Joseph E. Gagnon. "The New US Currency Policy." Peterson

Institute for International Economics, April 29, 2016. https://www
.piie.com/blogs/realtime-economic-issues-watch/new-us-currency-policy.

Blitz, Roger. "Countries Vie for Cryptocurrency Supremacy as Libra Tips Scales." *Financial Times,* December 12, 2019.

"A Bold Economic Switch." *Life,* August 27, 1971.

Bonafede, Dom. "Peterson Unit Helps Shape Tough International Economic Policy." *National Journal* 3, no. 46 (November 13, 1971): 2238–48.

Broder, David S. "Nixon Wins With 290 Electoral Votes." *Washington Post,* November 7, 1968.

Brzozowski, Alexandra. "Six European Countries Join EU-Iran Financial Trading Mechanism INSTEX." *Euractiv* (blog), November 29, 2019. https://www.euractiv.com/section/global-europe/news/six-european-countries-join-eu-iran-financial-trading-mechanism-instex/.

Burck, Gilbert. "Hard Going for the Game Plan." *Fortune,* May 1970.

Burka, Paul. "The Truth about John Connally." *Texas Monthly,* November 1979. https://www.texasmonthly.com/politics/the-truth-about-john-connally/.

"Business: Japan, Inc.: Winning the Most Important Battle." *Time,* May 10, 1971. http://content.time.com/time/subscriber/article/0,33009,902974,00.html.

"Call to Economic Revival." *New York Times,* August 16, 1971, sec. Editorial.

Camdessus, Michael, and Alexandre Lamfalussy. "Palais Royal Initiative-Reform of the International Monetary System: A Cooperative Approach for the 21st Century, February 8, 2011." In Michael Camdessus and Alexandre Lamfalussy, *Reform of the International Monetary System: The Palais Royal Initiative.* Thousand Oaks, CA: Sage, 2011.

Cameron, Juan. "How the U.S. Got on the Road to a Controlled Economy." *Fortune,* January 1972.

———. "The Last Reel of 'Mr. Peterson Goes to Washington.'" *Fortune,* March 1973.

———. "Richard Nixon's Very Personal White House." *Fortune,* July 1970.

Chan, Sewell. "Paul W. McCracken, Adviser to Presidents, Dies at 96." *New York Times,* August 4, 2012, sec. Business.

Collier, Barnard Law. "The Road to Peking, or, How Does This Kissinger Do It?" *New York Times Magazine,* November 14, 1971.

"Connally Sees Hope for Early Settlement of Monetary Problems, Perhaps by Jan. 1." *Wall Street Journal,* December 3, 1971.

"Connally's Hard Sell Against Inflation." *BusinessWeek,* no. 2184 (July 10, 1971): 62–66.

Cooper, Helene. "Plan to Cut U.S. Troops in West Africa Draws Criticism from Europe." *New York Times*, January 14, 2020, sec. World.

Cooper, Richard N. "Trade Policy Is Foreign Policy." *Foreign Policy*, no. 9 (1972): 18–36. https://doi.org/10.2307/1148083.

Cray, Douglas W. "Pact Encourages Business Experts." *New York Times*, December 19, 1971.

Dale, Edwin L., Jr. "David Kennedy Hedges on $35 Gold." *New York Times*, December 18, 1968.

———. "Economists Urge List of Reforms: 12 Insist World Needs More Than Monetary Changes." *New York Times*, December 22, 1971.

———. "Everyone Is Angry at the U.S. as I.M.F. Talks Begin." *New York Times*, September 26, 1971.

———. "Monetary Fund Approves World Currency Moves." *New York Times*, December 20, 1971.

———. "Monetary Talks Make Slow Start: Group of 10 Finance Chiefs Meet—Connally Is Hopeful." *New York Times*, December 18, 1971.

———. "Nixon Is Hopeful on Money Talks." *New York Times*, November 25, 1971.

———. "U.S. Lists Another Huge Payments Gap." *New York Times*, December 17, 1971.

Dale, Reginald. "Brussels Welcomes News." *Financial Times*, December 15, 1971.

Davie, Michael. "Man Who Went to Peking." *The Observer (1901–2003); London (UK)*, July 18, 1971.

Denevi, Timothy. "The Striking Contradictions of Richard Nixon's Inauguration 50 Years Ago, as Observed by Hunter S. Thompson." *Time*, January 19, 2019. https://time.com/5506809/richard-nixon-inauguration/.

Derby, Michael S. "Powell Says Fed Has No Plans to Create Digital Currency." *Wall Street Journal*, November 21, 2020, sec. U.S. https://www.wsj.com/articles/feds-powell-says-in-letter-to-congress-fed-not-creating-digital-currency-11574356188.

de Vries, Tom. "Jamaica, or the Non-Reform of the International Monetary System." *Foreign Affairs* 54, no. 3 (April 1976).

Dicks, Adrian. "U. S. Must Abide by Its 'Moral Undertakings,' Says Pompidou." *Financial Times*, December 23, 1971.

"The Dollar: A Power Play Unfolds." *Time*, August 30, 1971.

"The Dollar Accord." *Washington Post*, December 21, 1971, sec. Editorial.

Dorfman, Dan. "Heard on the Street." *Wall Street Journal*, August 11, 1971.

"The Drastic Plan to Save the Dollar." *BusinessWeek*, August 21, 1971.

Drummond, Roscoe. "Nixon Appointments Aim to Win Support of Center." *Washington Post*, December 18, 1968.

———. "Where Is John Connally's Future?" *Christian Science Monitor*, July 24, 1971.

"Economic Adviser to U.S. Presidents." *Washington Post*, August 6, 2012.

"The Economy: The Pragmatic Professor." *Time*, March 3, 1961. http://content.time.com/time/subscriber/article/0,33009,897654,00.html.

"The Economy: We Are All Keynesians Now." *Time*, December 31, 1965. http://content.time.com/time/subscriber/article/0,33009,842353,00.html.

Egan, Jack. "Stock, Dollar, Bond Markets Soar; Dow Up 35." *Washington Post*, November 2, 1978.

Eichengreen, Barry. "The Dollar and Its Discontents." Project Syndicate, October 18, 2018.

"Europe and the Dollar Problem." *Financial Times*, June 1, 1971.

"Exploring the New Economic World." *Time* 98, no. 9 (August 30, 1971): 10–13.

Farnsworth, Clyde H. "Dollar Challenged." *New York Times*, May 28, 1971.

———. "Dollar Sags Sharply as Nations Move Closer to Monetary Pact." *New York Times*, December 16, 1971.

———. "Economic Ministers Are Facing a Rough Monetary Confrontation in the Group of 10 Meeting in London." *New York Times*, September 15, 1971.

———. "European Hopes Buoyed." *New York Times*, December 19, 1971.

———. "Europeans Apply Dollar Pressure." *New York Times*, May 12, 1971.

———. "Gold Price Rises to a 2▨Year High." *New York Times*, July 28, 1971. https://www.nytimes.com/1971/07/28/archives/gold-price-rises-to-a-2year-high-demand-in-europe-markets-grows.html.

———. "Group of 10, in Rome, Begins Negotiating Realignment of Currencies." *New York Times*, December 1, 1971.

———. "Health of Dollar Stirs New Worry." *New York Times*, July 30, 1971. https://www.nytimes.com/1971/07/30/archives/health-of-dollar-stirs-new-worry-gold-sliver-and-the-strong.html.

———. "Hope Stirs as Group of 10 Gathers for Meeting." *New York Times*, November 29, 1971, sec. Business/Finance.

———. "Monetary Breach Held Narrowing." *New York Times*, December 2, 1971.

———. "O.E.C.D. Experts Express Caution." *New York Times*, December 23, 1971.

———. "Tell International Forum How Overseas Dollar Market Operates." *New York Times*, May 27, 1971.

———. "U.S. Ready to End Surtax If Currencies Go Up 11%." *New York Times*, November 30, 1971.

"Foreign Reaction to Dollar Defense." *Wall Street Journal*, November 3, 1978.

Foroohar, Rana. "The Allure of Financial Tricks Is Fading." *Financial Times*, March 3, 2019.

———. "American Capitalism's Great Crisis." *Time*, May 12, 2016. https://time.com/4327419/american-capitalisms-great-crisis/.

Frankel, Jeffrey. "The Plaza Accord 30 Years Later." *International Monetary Cooperation: Lessons from the Plaza Accord After Thirty Years*. Washington, DC: Peterson Institute for International Economics, 2016, 53–72.

Frankel, Max. "Behind the Speech: Some Principal Actors in Turnover Are Unseen at Inaugural Spectacle." *New York Times*, January 21, 1969.

Frankel, Max, and Robert B. Semple Jr. "If It's Thursday, This Must Be Rashtrapati Bhavan." *New York Times Magazine*, August 17, 1969.

"The Free World Has Won." *New York Times*, December 20, 1971, sec. Editorial.

Freeland, Chrystia. "Lunch with the FT: Pete Peterson." *Financial Times*, January 26, 2007.

"The Freeze Starts with a Fever." *BusinessWeek*, August 21, 1971.

Friedman, Milton. "Economic Perspective." *Newsweek*, December 22, 1969.

Gagnon, Joseph E. "The Unsustainable Trajectory of US International Debt." Peterson Institute for International Economics, March 29, 2017. https://www.piie.com/blogs/realtime-economic-issues-watch/unsustainable-trajectory-us-international-debt.

Garten, Jeffrey E. "Gunboat Economics." *Foreign Affairs* 63, no. 3 (1984): 538–59. https://www.jstor.org/stable/20042271.

———. "Lessons for the Next Financial Crisis." *Foreign Affairs* 78, no. 2 (April 1999): 76–92. https://www.jstor.org/stable/20049210.

———. "The 100-Day Economic Agenda." *Foreign Affairs* 71, no. 5 (Winter 1992): 16–31. https://www.jstor.org/stable/20045400.

Georgieva, Kristina. "A New Bretton Woods Moment." IMF Annual Meeting 2020. October 15, 2020. https://www.imf.org/en/News/Articles/2020/10/15/sp101520-a-new-bretton-woods-moment.

Giorgione, Michael. "Inside Camp David," article excerpt, *The National*, n.d., http://www.amtrakthenational.com/inside-camp-david.

"Gold Outglitters the West German Mark on World Markets; Speculators Sit Tight." *Wall Street Journal*, May 13, 1971.

Goodman, Peter S. "The Dollar Is Still King. How (in the World) Did That Happen?" *New York Times*, February 22, 2019, sec. Business.

Gopinath, Gita. "Digital Currencies Will Not Displace the Dominant Dollar." *Financial Times*, January 7, 2020.

Grose, Peter. "Kissinger Gains a Key Authority in Foreign Policy." *New York Times*, February 5, 1969.

"Group of 10 Fails to Win Monetary Accord as U.S. Holds Out for Full Nixon Package." *Wall Street Journal*, September 17, 1971.

Gup, Ted. "Underground Government." *Washington Post*, May 31, 1992.

Hajric, Vildana. "Davos Digital-Payments Rift Pits Facebook Against Central Banks." Bloomberg.com, January 23, 2020.

Harding, Robin. "G20 Convenes in Japan with Currency Wars the Next Big Worry." *Financial Times*, June 27, 2019.

Heinemann, H. Erich. "Chaotic Trading Weakens the Dollar." *New York Times*, August 14, 1971.

———. "The Dollar as Kingpin Is Dead: Any New Monetary System Will Give It Smaller Role." *New York Times*, December 20, 1971.

Herbers, John. "The 37th President; In Three Decades, Nixon Tasted Crisis and Defeat, Victory, Ruin and Revival." *New York Times*, April 24, 1994, sec. A.

Hershey, Robert D., Jr. "Peter G. Peterson, a Power from Wall St. to Washington, Dies at 91." *New York Times*, March 20, 2018, sec. Obituaries.

Hinane El Kadi, Tin. "The Promise and Peril of the Digital Silk Road." Chatham House, June 6, 2019. https://www.chathamhouse.org/expert /comment/promise-and-peril-digital-silk-road.

"History of FX at CME Group." CME Group, January 24, 2020. http:// www.cmegroup.com/trading/fx/fxhistory.html.

"How Burns Will Change the Fed." *BusinessWeek*, October 25, 1969.

Huebner, Lee. "The Checkers Speech After 60 Years." *The Atlantic*, September 22, 2012. https://www.theatlantic.com/politics/archive/2012/09/ the-checkers-speech-after-60-years/262172/.

Humpage, Owen. "The Smithsonian Agreement." federalreservehistory.org, November 22, 2013. https://www.federalreservehistory.org/essays/smith sonian_agreement.

Hunt, Albert R. "Leaders' Lament: Business Council's Inflation Gripes May Renew Nixon Leadership Issue." *Wall Street Journal*. October 19, 1970.

"In Search of a New Balance." *Financial Times*, August 17, 1971, sec. Editorial.

Irwin, Douglas A. "The Nixon Shock After Forty Years: The Import Surcharge Revisited." *World Trade Review* 12, no. 1 (January 2013): 29–56.

Irwin, Neil. "Paul A. Volcker, Fed Chairman Who Curbed Inflation by Raising Interest Rates, Dies at 92." *Washington Post*, December 9, 2019.

Isaac, Anna, and Caitlin Ostroff. "Central Banks Warm to Issuing Digital Currencies." *Wall Street Journal*, January 23, 2020, sec. Markets.

Janssen, Richard F. "Dollar Devaluation: It Could Be Tricky." *Wall Street Journal*, December 15, 1971.

———. "Monetary Parley Opening in Washington Amid Hints Parities Dispute Is Near End." *Wall Street Journal*, December 17, 1971.

———. "The New Dollar: Devaluation of 8.57% Likely to Create Jobs, Help Nixon Summitry." *Wall Street Journal*. December 20, 1971.

———. "New Treasury Chief: Choice of Connally Is Seen Aimed at Congress Ties, '72 Texas Vote." *Wall Street Journal*, December 15, 1970.

———. "Nixon Agrees to Formally Devalue Dollar in Azores Announcement with Pompidou." *Wall Street Journal*, December 15, 1971.

Janssen, Richard F., and Albert R. Hunt. "War of Nerves Quickens: White House Hints It Plans Attack on Reserve Board's Independence." *Wall Street Journal*, July 29, 1971.

Johnson, James A. "The New Generation of Isolationists." *Foreign Affairs* 49, no. 1 (October 1970): 136–46. https://www.jstor.org/stable/2003 7824.

Kagan, Robert. "The Cost of American Retreat." *Wall Street Journal*, September 7, 2018, sec. Life.

Keegan, William. "Group of Ten Talks End in Gridlock." *Financial Times*, September 17, 1971.

Kharpal, Arjun. "Calls for a US 'Digital Dollar' Rise as China Powers Ahead with a Digital Yuan." CNBC, January 23, 2020. https://www.cnbc .com/2020/01/23/davos-calls-for-a-us-digital-dollar-as-china-works-on -digital-yuan.html.

Kida, Kazuhiro, Masayuki Kubota, and Yusho Cho. "Rise of the Yuan: China-Based Payment Settlements Jump 80%." *Nikkei Asian Review*, May 20, 2019. https://asia.nikkei.com/Business/Markets/Rise-of-the -yuan-China-based-payment-settlements-jump-80.

"Kissinger: The Uses and Limits of Power." *Time*, February 14, 1969. http:// content.time.com/time/subscriber/article/0,33009,900610,00.html.

Kristof, Nicholas D. "Dollar Plunges to 16-Month Low in Reaction to 5 Nations' Accord." *New York Times*, September 24, 1985.

Lee, John M. "$13-Billion Gain Sought to Spur Payments to U.S." *New York Times*, September 16, 1971.

———. "Group of 10 Fails to Find Accord on Dollar Crisis." *New York Times*, September 17, 1971.

Lehrman, Lewis E. "The Nixon Shock Heard 'Round the World." *Wall Street*

Journal, August 15, 2011, sec. Opinion. https://www.wsj.com/articles/SB 10001424053111904007304576494073418802358.

Levine, Richard J. "Campus to Cabinet: Shultz, an Ex-Professor, Wins General Acclaim as Secretary of Labor." *Wall Street Journal*, June 9, 1969.

Lewis, Paul. "Economists Propose Blue-Print." *Financial Times*, December 23, 1971.

———. "Kissinger Optimistic for Monetary Settlement." *Financial Times*, December 1, 1971.

———. "Nixon Agrees to a Dollar Devaluation." *Financial Times*, December 15, 1971.

———. "U.S. Monetary Views More Flexible." *Financial Times*, November 29, 1971.

Lowenstein, Roger. "The Nixon Shock." Bloomberg.com, August 4, 2011.

Malabre, Alfred L., Jr. "Idea of Floating Rates Of Exchange Is Gaining, But Misgivings Persist." *Wall Street Journal*, July 12, 1968.

———. "Monetary Crisis Cools but Could Well Reheat; Ill Will Grows Abroad." *Wall Street Journal*, May 21, 1971.

Malkin, Lawrence. "A Practical Politician at the Fed." *Fortune* 83 (May 1971): 148–51, 254, 259–60, 262, 264.

Marshall, Jon. "Nixon Is Gone, but His Media Strategy Lives On." *The Atlantic*, August 4, 2014. https://www.theatlantic.com/politics/archive/2014/08/nixons-revenge-his-media-strategy-triumphs-40-years-after-resignation/375274/.

McCracken, Paul W. "Economic Policy in the Nixon Years." *Presidential Studies Quarterly* 26, no. 1 (1996): 165–77. https://www.jstor.org/stable/27551556.

McCulloch, Rachel. "Unexpected Real Consequences of Floating Exchange Rates." Vol. 153, "Essays in International Finance." Princeton, NJ: Princeton University Press, 1983. https://doi.org/10.1007/978–1–349–18513–9_3.

McWhirter, William A. "'An Expert Views the Freeze and the Future': An Interview of Otto Eckstein." *Life*, August 27, 1971.

Mossberg, Walter. "UAW's GM Unit Clears Contract by a 4–1 Margin." *Wall Street Journal*. November 13, 1970, sec. 1.

"Mr. Peterson's Assignment." *Fortune*, March 1971.

"NATO Members' Promise of Spending 2% of Their GDP on Defence Is Proving Hard to Keep." *The Economist*, March 14, 2019.

Naughton, James M. "Shultz Quietly Builds Up Power in Domestic Field." *New York Times*, May 31, 1971.

"The New Activist in Central Banking." *BusinessWeek*, no. 2193 (September 11, 1971): 120.

"The New Economic Policy." *Fortune*, September 1971.

"The New Lineup: Devaluation of Dollar, Revaluation of Others Seen Aiding Economy." *Wall Street Journal*, December 15, 1971.

"New Man at the Treasury." *New York Times*, December 15, 1970, sec. Editorial Board.

"New Perspectives on Trade." *Wall Street Journal*, August 20, 1971.

"Nixon Is Shifting to a Harder-Hitting Game." *BusinessWeek*, no. 2154 (December 12, 1970): 14–15.

"Nixon 'Steady-as-You-Go' Economic Policy Is Proper, Declares Budget Chief Shultz." *Wall Street Journal*, April 23, 1971.

"Nixon's Go-Go Economic Policies for '72." *BusinessWeek*, January 29, 1972.

"Nixon's Grand Design for Recovery." *Time*, August 30, 1971.

"Nixon's New Summitry." *Financial Times*, November 30, 1971, sec. Editorial.

"Now for the Grind." *Wall Street Journal*, November 3, 1978, sec. Editorial.

Oberdorfer, Don. "Connally Selection Seen as Bid for Texas in 1972." *Washington Post*, December 15, 1970.

———. "Nixon's Economic Advisers Called to Weekend Sessions," *Washington Post*, August 14, 1971.

Oberdorfer, Don, and Frank C. Porter. "Connally Urges Tough Trade Stance." *Washington Post*, April 26, 1971.

Obstfeld, Maurice, and Alan M. Taylor. "International Monetary Relations: Taking Finance Seriously." *Journal of Economic Perspectives* 31, no. 3 (2017): 3–28. https://www.jstor.org/stable/44321277.

Odell, John S. "Going Off Gold and Forcing Dollar Depreciation." *U.S International Monetary Policy: Markets, Power, and Ideas as Sources of Change*. Princeton, NJ: Princeton University Press, 2014.

"On Devaluation." *Washington Post*, December 15, 1971, sec. Editorial.

Osipovich, Alexander. "Former Regulator Known as 'Crypto Dad' to Launch Digital-Dollar Think Tank." *Wall Street Journal*, January 16, 2020, sec. Markets. https://www.wsj.com/articles/former-regulator-known-as-crypto-dad-to-launch-digital-dollar-think-tank-11579179604.

Pierce, James L. "The Political Economy of Arthur Burns." *Journal of Finance* 34, no. 2 (1979): 485–96. https://doi.org/10.2307/2326989.

Pierson, John. "Jobs Bill Vetoed: Nixon Is Satisfied Policies to Cut Joblessness, Inflation Will Progress." *Wall Street Journal*, June 30, 1971, sec. 1.

———. "Trade Tightrope: Nixon Aide Peterson Has Controversial Ideas on Overseas Dealings." *Wall Street Journal*, July 6, 1971.

———. "White House Power: Arthur Burns Provides Conservative Influence on Domestic Programs." *Wall Street Journal*, May 20, 1969, sec. 1.

———. "Who's in Charge? Coming Cabinet Moves Point Up the Big Decline of Secretaries' Role." *Wall Street Journal*, November 20, 1970.

Pine, Art. "Behind the Decision: Far-Reaching Dollar Rescue Program Formulated by Top Officials in Secrecy." *Washington Post*, November 2, 1978.

"President at Camp David Preparing Summit Talks." *Washington Post*, December 12, 1971.

"President Nixon Takes Off the Kid Gloves." *BusinessWeek*, August 21, 1971.

"Price of Gold Soars on Global Markets Following U.S. Report." *Wall Street Journal*, July 28, 1971, sec. 1.

"Problems of Attitude." *Wall Street Journal*, August 17, 1971.

Rediker, Douglas A. "Why US Multilateral Leadership Was Key to the Global Financial Crisis Response." *Brookings* (blog), September 12, 2018. https://www.brookings.edu/blog/future-development/2018/09/12/why-us-multilateral-leadership-was-key-to-the-global-financial-crisis-response/.

Reston, James. "From Promise to Policy: Many Pitfalls Await Efforts by Nixon to Redeem Pledges and Unify Nation." *New York Times*, November 7, 1968.

———. "Kissinger: New Man in the White House Basement." *New York Times*, December 4, 1968.

———. "A Remarkable Comeback for Nixon." *New York Times*, August 9, 1968.

———. "60's Close with Some Hope for Peace at Home and Abroad." *New York Times*, December 30, 1969, sec. Special Supplement.

Ripley, S. Dillon. "The View from the Castle." *Smithsonian Magazine*, February 1972.

Robards, Terry. "Currency Accord Sends Stocks Up in Heavy Trading." *New York Times*, December 21, 1971.

———. "Elections Viewed Lacking as an Economic Mandate." *New York Times*, November 5, 1970, sec. Business & Finance.

Rogaly, Joe. "Gold Price Change Not Anticipated Says Nixon's Aide." *Financial Times*, December 19, 1968.

Rowen, Hobart. "David Kennedy Offers Views: Nixon Held Flexible on Gold Price Rise." *Washington Post*, December 18, 1968.

———. "Devaluation: Presidential Midas Touch: News Analysis." *Washington Post*, December 20, 1971.

———. "General Enthusiasm Greets Choice of McCracken." *Washington Post*, December 5, 1968.

———. "Group of 10 Envisions Money Agreement by Weekend." *Washington Post*, December 16, 1971, sec. General.

———. "Monetary Agreement Reached: U.S. Devalues; Nixon Unveils Historic Pact." *Washington Post*, December 19, 1971, sec. General.

———. "New Patterns for Money: President Abandons Historic Stand." *Washington Post*, December 15, 1971.

———. "Profiles of the Nixon Cabinet: David M. Kennedy." *Washington Post,* December 12, 1968.

———. "U.S. Going Protectionist Following Monetary Crisis." *Washington Post*, June 7, 1971.

———. "West, Japan Pressured on Trade, Arms." *Washington Post*, May 29, 1971.

Rugaber, Walter. "37th U.S. President: Leader Casting Many Images." *New York Times*, January 21, 1969.

Samuelson, Paul A. "Bleak Outlook." *Newsweek*, March 2, 1970.

———. "Gold." *Newsweek*, October 14, 1968.

———. "The New Economics." *Newsweek*, November 25, 1968.

Samuelson, Robert. "The Decade of Retreat." *Washington Post*, December 26, 2019. https://www.washingtonpost.com/opinions/2019/12/26/s-were-decade-what-exactly-six-columnists-tell-us/.

Sandbu, Martin. "Europe First: Taking on the Dominance of the US Dollar." *Financial Times*, December 5, 2019.

Segal, Troy. "Forex Market: Who Trades Currency and Why." Investopedia, October 24, 2019. https://www.investopedia.com/articles/forex/11/who-trades-forex-and-why.asp.

Semple, Robert B., Jr. "Nixon, Sworn, Dedicates Office to Peace." *New York Times*, January 21, 1969.

Sensit, Michael R., Stephen Grover, and Bernard Wysocki Jr. "Some Executives Are Skeptical of Steps to Lower Dollar." *Wall Street Journal*, September 23, 1985, sec. 1.

Shribman, David. "Congress Is Relieved by Efforts on Dollar, But Some Stress Need to See Fast Results." *Wall Street Journal*, September 23, 1985, sec. 1.

Sidey, Hugh. "The Economic Bombshell." *Life*, August 27, 1971.

Silk, Leonard. "The Accord of 1970." *New York Times*, December 9, 1970, sec. Archives. https://www.nytimes.com/1970/12/09/archives/the-accord-of-1970-speeches-by-nixon-and-burns-show-approach-on.html.

———. "Dollar Devaluation." *New York Times*, September 29, 1971.

———. "Economic Poker Game: Americans Playing for Major Stakes as

Group of 10 Convenes in London." *New York Times*, September 15, 1971, sec. Business/Financial.

———. "New Realities Abroad." *New York Times*, March 8, 1972.

———. "Shaping World Economy: A Managed System of Floating Rates Appears to Be I.M.F.'s Accepted Plan." *New York Times*, January 7, 1976.

———. "What Now for Dollar?" *New York Times*, December 22, 1971.

Smith, William D. "U.S. Economists Hail Monetary Accord." *New York Times*, December 20, 1971, sec. Business/Financial.

Sommer, Jeff. "Why Tariff and Trade Disputes Are More than a Money Problem." *New York Times*, September 21, 2018. https://www.nytimes.com/2018/09/21/business/trade-disputes-more-than-a-money-problem.html.

"Squeeze on Japan." *New York Times*, August 20, 1971, sec. Editorial.

Stetson, Damon. "Unions Reject No-Strike Appeal." *New York Times*, August 19, 1971. https://www.nytimes.com/1971/08/19/archives/unions-reject-nostrike-appeal-uaw-chief-irate-labor-leader-warns.html.

Stout, David. "Peter G. Peterson, Financier Who Warned of Rising National Debt, Dies at 91." *Washington Post*, March 20, 2018, sec. Obituaries.

Strange, Susan. "The Dollar Crisis 1971." *International Affairs* 48, no. 2 (1972): 191–216. https://doi.org/10.2307/2613437.

———. "The Unsettled System: Stockholm to the Smithsonian." *International Economic Relations of the Western World, 1959–1971*. Vol. 2, *International Monetary Relations*, 320–53. London: University of Oxford Press, 1976.

"Support Abroad: Foreign Officials Supporting Carter's Action on Dollar." *Washington Post*, November 2, 1978.

Szalay, Eva, and Hudson Lockett. "Why the Renminbi's Challenge to the Dollar Has Faded." *Financial Times*, October 15, 2019.

Szulc, Tad. "Letter from the Azores." *The New Yorker*, January 1, 1972.

———. "Nixon Agrees to Devaluation of Dollar as Part of Revision of Major Currencies." *New York Times*, December 15, 1971.

Tankersley, Jim. "Budget Deficit Topped $1 Trillion in 2019." *New York Times*, January 13, 2020, sec. Business.

Tether, C. Gordon. "$—When Something Has Got to Give." *Financial Times*, August 10, 1971.

"Text of Blumenthal-Miller, Carter Statements: Administration Says That 'The Time Has Come to Call a Halt to This Development.'" *Washington Post*, November 2, 1978.

"Text of Connally Talk to Rich Nations in London." *Washington Post*, September 26, 1971, sec. Business & Finance.

Tooze, Adam. "The Forgotten History of the Financial Crisis." *Foreign Affairs* 97, no. 5 (October 2018): 199–210.

Truman, Edwin M. "The End of the Bretton Woods International Monetary System." Working Paper. Peterson Institute for International Economics, October 2017. https://www.piie.com/publications/working-papers/end-bretton-woods-international-monetary-system.

"U.S. Gold Stock Dropped in May By \$357 Million." *Wall Street Journal*, June 24, 1971, sec. 1.

"The U.S. Searches for a Realistic Trade Policy." *BusinessWeek*, no. 2183 (July 3, 1971): 64–70.

"Very Like a Whale." *Wall Street Journal*, December 20, 1971, sec. Editorial.

Viorst, Milton. "The Burns Kind of Liberal Conservatism." *New York Times Magazine*, November 9, 1969.

Wadhams, Nick, and Jennifer Jacobs. "Trump Seeks Huge Premium from Allies Hosting U.S. Troops." Bloomberg.com, March 8, 2019.

"Washington Agreement." *Financial Times*, December 20, 1971, sec. Editorial.

"Weakening the Dollar Is Not Enough." *New York Times*, September 24, 1985.

"The Weather." *Frederick News Post*, August 14, 1971.

Whalen, Richard. "The Nixon-Connally Arrangement." *Harper's*, August 1971.

Winters, Bruce. "Shultz Noted as A Skilled, Cool Labor Secretary." *The Baltimore Sun*, June 11, 1970.

Wolf, Martin. "Martin Wolf on Bretton Woods at 75: Global Co-Operation Under Threat." *Financial Times*, July 10, 2019.

Official Documents and Primary Materials
(Includes reports, hearings, memos, diary entries, dissertations and theses, speeches, statements, oral histories)

"Action Memorandum from the President's Assistant for National Security Affairs (Kissinger) and the President's Assistant for International Economic Affairs (Peterson) to President Nixon, September 20, 1971. Subject: Negotiating the New Economic Policy Abroad." Document 176 in *Foreign Relations of the United States, 1969–1976*. Vol. 3, *Foreign Economic Policy; International Monetary Policy, 1969–1972*. Washington, DC: U.S. Government Printing Office, 2001. https://history.state.gov/historicaldocuments/frus1969–76v03/d176.

"Action Now to Strengthen the U.S. Dollar: Report of the Subcommittee on International Exchanges and Payments of the Joint Economic Committee, Congress of the United States," August 1971, Washington, DC: U.S. Government Printing Office, 1971. ProQuest Congressional.

Auten, John. "Memorandum from John Auten, Office of Financial Analysis, to Paul Volcker, Regarding U.S. Exchange Rate, May 28, 1971." Box 108477. Volcker Papers, New York Federal Reserve Archive.

Author's In-Person Tour of the Presidential Helicopter. August 14, 2019. Richard M. Nixon Presidential Library and Museum, Yorba Linda, CA.

Bank for International Settlements. "OTC Derivatives Statistics at End-June 2018," October 31, 2018. https://www.bis.org/publ/otc_hy1810.htm.

Bergsten, Fred C. "Information Memorandum from C. Fred Bergsten of the National Security Council Staff to the President's Assistant for National Security Affairs (Kissinger), April 14, 1969." Document 19 in *Foreign Relations of the United States, 1969–1976*. Vol. 3, *Foreign Economic Policy; International Monetary Policy, 1969–1972*. Washington, DC: U.S. Government Printing Office, 2001. https://history.state.gov/historical documents/frus1969–76v03/d19.

Burns, Arthur F. "The Basis for Lasting Prosperity, Address by Arthur F. Burns in the Pepperdine College Great Issues Series, December 7, 1970." In his *Reflections of an Economic Policy Maker: Speeches, Congressional Statements: 1969–1978*, 103–15. Washington, DC: American Enterprise Institute, 1978.

———. "Inflation: The Fundamental Challenge to Stabilization Policies: Remarks Before the Seventeenth Annual Monetary Conference of the American Bankers Association, Hot Springs, Virginia, May 18, 1970." In *Reflections of an Economic Policy Maker: Speeches, Congressional Statements: 1969–1978*, 91–102. Washington, DC: American Enterprise Institute, 1978.

———. "Notes on Camp David Weekend," n.d. Provided to author by Josh Zoffer, Yale Law School.

———. "Remarks of the President and Dr. Arthur F. Burns at the Swearing In of Dr. Burns as Chairman, January 31, 1970." In *Weekly Compilation of Presidential Documents* 6, no. 5 (January 31, 1970): 97–98. https://hdl .handle.net/2027/mdp.39015087529395?urlappend=%3Bseq=96.

———. "Statement Before the Joint Committee of the U.S. Congress, July 23, 1971 ('The Economy in Mid-1971')." In his *Reflections of an Economic Policy Maker: Speeches, Congressional Statements: 1969–1978*, 117–27. Washington, DC: American Enterprise Institute, 1978.

Cecchetti, Stephen G., and Enisse Kharroubi. "Reassessing the Impact of Finance on Growth." BIS Working Paper No. 381, July 2012, Bank for

International Settlements, Basel, Switzerland. https://www.bis.org/publ/work381.pdf.

"Central Bank Group to Assess Potential Cases for Central Bank Digital Currencies," January 21, 2020. https://www.bis.org/press/p200121.htm.

Commission on the Role of Gold in the Domestic and International Monetary Systems. "Report to the Congress of the Commission on the Role of Gold in the Domestic and International Monetary Systems, Volume I," March 31, 1982, Washington, DC. https://fraser.stlouisfed.org/title/339#6346.

Connally, John B. "Inaugural Address, January 15, 1963." In *Journal of the House of Representatives of the Regular Session of the Fifty-Eighth Legislature of the State of Texas,* n.d. https://lrl.texas.gov/legeLeaders/governors/displayDocs.cfm?govdoctypeID=6&governorID=37.

———. "Memorandum for Alexander P. Butterfield" with draft for Richard Nixon for IMF Remarks, September 22, 1971. Papers of John B. Connally, 77–353B: JBC Official Chronology, September–November 1971, Lyndon Baines Johnson Presidential Library, Austin, TX.

———. "Memorandum for the President Prepared by John Connally, 'Monetary and Trade Issues Aiming at the Azores Meeting,' December 10, 1971." Papers of John B. Connally, 77–353C: JBC–Official Chronology–December 1971. Lyndon Baines Johnson Presidential Library. Austin, TX.

———. "Memorandum from Secretary of the Treasury Connally to President Nixon, June 8, 1971." Document 158 in *Foreign Relations of the United States, 1969–1976.* Vol. 3, *Foreign Economic Policy; International Monetary Policy, 1969–1972.* Washington, DC: U.S. Government Printing Office, 2001. https://history.state.gov/historicaldocuments/frus1969–76v03/d158.

———. "Mutual Responsibility for Maintaining a Stable Monetary System, Address by John B. Connally Before the American Bankers Association at Munich, Germany on May 28." In *Department of State Bulletin July 12, 1971,* 42–46, 1971. https://hdl.handle.net/2027/msu.31293008122040?urlappend=%3Bseq=48.

———. "News Conference of Secretary John B. Connally Following Meetings in the Azores with Finance Minister Giscard D'Estaing, December 13, 1971." *Weekly Compilation of Presidential Documents* 7, no. 51 (December 20, 1971): 1658–61. https://hdl.handle.net/2027/osu.32437011290398?urlappend=%3Bseq=716.

———. "Remarks of Secretary of the Treasury John B. Connally, Jr., at a Press Conference Following a Series of Discussions by Administration

Officials on Economic and Budget Matters, June 29, 1971." In *Weekly Compilation of Presidential Documents* 7, No. 27:1002–4. Washington, DC: U.S. Government Printing Office, 1971. https://hdl.handle.net /2027/osu.32437011290398?urlappend=%3Bseq=10.

———. "Secretary Connally's Statement at G-10 Meeting." September 15, 1971. Papers of John B. Connally, 49–191C: G-10 Meetings—London 9/15, 16, 17/71, Lyndon Baines Johnson Presidential Library, Austin, TX.

———. "Statement by Secretary of the Treasury John B. Connally at the Opening of a News Conference, August 16, 1971," Papers of John B. Connally, 49–191C, Press Conference—Major Economic Programs, Monday, 8/16/71. Lyndon Baines Johnson Presidential Library, Austin, TX.

———. "Statement by the Governor of the Fund and Bank for the United States." In *International Monetary Fund Summary Proceedings of the Twenty-Sixth Annual Meeting of the Board of Governors, September 27– October 1, 1971,* 1971.

Connally, John M., George P. Shultz, and Paul W. McCracken. "Press Conference of Hon. John M. [*sic*] Connally, Secretary of the Treasury; Hon. George P. Shultz, Director, Office of Management and Budget; and Hon. Paul W. McCracken, Chairman, Council of Economic Advisers," August 15, 1971. FRC Box 7, Records of Secretary of the Treasury George P. Shultz, 1971– 1974, Record Group 56, National Archives, College Park, MD.

Currency Manipulation and Its Toll on the US Economy. Peterson Institute for International Economics, 2017. https://www.youtube.com /watch?v=wE7UkXk_VCo.

De Gregorio, José, Barry Eichengreen, Takatoshi Itō, and Charles Wyplosz. *IMF Reform: The Unfinished Agenda.* Geneva Reports on the World Economy 20. Geneva: International Center for Monetary and Banking Studies, 2018. https://cepr.org/sites/default/files/events/Geneva20.pdf.

Dornbusch, Rudiger, and Jeffrey A. Frankel. "The Flexible Exchange Rate System: Experience and Alternatives." Working Paper. National Bureau of Economic Research, December 1987. https://doi.org/10.3386/w2464.

"Editorial Note." Document 5 in *Foreign Relations of the United States, 1969– 1976.* Vol. 1, *Foundations of Foreign Policy, 1969–1972.* Washington, DC: U.S. Government Printing Office, 2003. https://history.state.gov /historicaldocuments/frus1969–76v01/d5.

"Editorial Note." Document 164 in *Foreign Relations of the United States, 1969–1976.* Vol. 3, *Foreign Economic Policy; International Monetary Policy, 1969–1972.* Washington, DC: U.S. Government Printing Office, 2001. https://history.state.gov/historicaldocuments/frus1969–76v03/d164.

"Editorial Note." Document 203 in *Foreign Relations of the United States, 1969–1976*. Vol. 3, *Foreign Economic Policy; International Monetary Policy, 1969–1972*. Washington, DC: U.S. Government Printing Office, 2001. https://history.state.gov/historicaldocuments/frus1969–76v03/d203.

"Editorial Note." Document 210 in *Foreign Relations of the United States, 1969–1976*. Vol. 3, *Foreign Economic Policy; International Monetary Policy, 1969–1972*. Washington, DC: U.S. Government Printing Office, 2001. https://history.state.gov/historicaldocuments/frus1969–76v03/d210.

"Editorial Note." Document 219 in *Foreign Relations of the United States, 1969–1976*. Vol. 3, *Foreign Economic Policy; International Monetary Policy, 1969–1972*. Washington, DC: U.S. Government Printing Office, 2001. https://history.state.gov/historicaldocuments/frus1969–76v03/d219.

"Editorial Note, March 2, 1970 (Nixon Memo to Haldeman on International Economic Policy)." Document 38 in *Foreign Relations of the United States, 1969–1976*. Vol. 3, *Foreign Economic Policy; International Monetary Policy, 1969–1972*. Washington, DC: U.S. Government Printing Office, 2001. https://history.state.gov/historicaldocuments/frus1969–76v03/d38.

Ehrlichman, John D. (Various documents), n.d. Box 6, Notes of Meeting with President, Staff Member Office Files, White House Special Files, Richard M. Nixon Presidential Library and Museum, Yorba Linda, CA.

Forbord, Thomas Austin. "The Abandonment of Bretton Woods: The Political Economy of U.S. International Monetary Policy." PhD diss., Harvard University, 1980.

"'Framework for Monetary and Trade Settlement,' Paper Agreed by President Nixon and President Pompidou." Document 220 in *Foreign Relations of the United States, 1969–1976*. Vol. 3, *Foreign Economic Policy; International Monetary Policy, 1969–1972*. Washington, DC: U.S. Government Printing Office, 2001. https://history.state.gov/historicaldocuments/frus1969–76v03/d220.

Garten, Jeffrey E., and Robert D. Hormats. "America's Future Role in the Global Economy: The U.S. Dollar." Unpublished Lecture. Presented at the Yale School of Management, New Haven, CT, December 11, 2019.

Gergen, David. An Oral History Interview with David Gergen. Interview by Timothy J. Naftali. Audio, August 5, 2009. Richard Nixon Presidential Library and Museum, Yorba Linda, CA.

German Embassy. "German Press Review," August 18, 1971. Richard M. Nixon Presidential Library and Museum, Yorba Linda, CA.

Gowa, Joanne. *Closing the Gold Window: Domestic Politics and the End of Bretton Woods*. Ithaca, NY: Cornell University Press, 1983.

————. "Explaining Large Scale Policy Change: Closing the Gold Window, 1971." PhD diss., Princeton University, 1980.

Graebner, Linda S. "The New Economic Policy, 1971." In *Appendices: Commission on the Organization of the Government for the Conduct of Foreign Policy*. Washington, DC, 1975, 160–84. http://hdl.handle.net/2027 /ucl.31158010972742.

"G7 Information Centre." http://www.g7.utoronto.ca/.

"G20 Information Centre." http://www.g20.utoronto.ca/.

Haberler, Gottfried. "Report of Task Force on U.S. Balance of Payments to the President-Elect." Unpublished, January 1969. Richard M. Nixon Presidential Library and Museum, Yorba Linda, CA.

Haldeman, H. R. "Diaries of H.R. Haldeman," 1969–1973. Haldeman Diary Transcripts, Richard Nixon Presidential Library and Museum, Yorba Linda, CA. https://www.nixonlibrary.gov/h-r-haldeman-diaries.

Hormats, Robert D. "Information Memorandum from Robert Hormats of the National Security Council Staff to the President's Assistant for National Security Affairs (Kissinger), August 13, 1971." Document 167 in *Foreign Relations of the United States, 1969–1976*. Vol. 3, *Foreign Economic Policy; International Monetary Policy, 1969–1972*. Washington, DC: U.S. Government Printing Office, 2001. https://history.state.gov /historicaldocuments/frus1969–76v03/d167.

————. "Information Memorandum from Robert Hormats of the National Security Council Staff to the President's Assistant for National Security Affairs (Kissinger), September 28, 1971. Subject: Progress in Developing USG Positions on International Aspects of NEP." Document 182 in *Foreign Relations of the United States, 1969–1976*. Vol. 3, *Foreign Economic Policy; International Monetary Policy, 1969–1972*. Washington, DC: U.S. Government Printing Office, 2001. https://history.state.gov /historicaldocuments/frus1969–76v03/d182.

————. "Information Memorandum from Robert Hormats of the National Security Council Staff to the President's Assistant for National Security Affairs (Kissinger), Subject: Talking Points for Your Meeting with George Shultz, Tuesday, November 2, at 8:45 a.m." Document 188 in *Foreign Relations of the United States, 1969–1976*. Vol. 3, *Foreign Economic Policy; International Monetary Policy, 1969–1972*. Washington, DC: U.S. Government Printing Office, 2001. https://history.state.gov /historicaldocuments/frus1969–76v03/d188.

International Aspects of the President's New Economic Policies: Hearings Before the Subcommittee on International Trade of the Committee on Finance. 92nd Cong., 1st Sess. (1971). ProQuest Congressional.

The International Implications of the New Economic Policy: Hearings Before the Subcommittee on Foreign Economic Policy of the Committee on Foreign Affairs, 92nd Cong., 1st Sess. (1971). ProQuest Congressional.

International Monetary Fund. *International Monetary Fund Summary Proceedings of the Twenty-Sixth Annual Meeting of the Board of Governors, September 27–October 1, 1971.* Washington, DC, 1971.

Johnston, Ernest. "Memorandum from Ernest Johnston of the National Security Council Staff to the President's Assistant for National Security Affairs (Kissinger), June 23, 1971." Document 160 in *Foreign Relations of the United States, 1969–1976.* Vol. 3, *Foreign Economic Policy; International Monetary Policy, 1969–1972.* Washington, DC: U.S. Government Printing Office, 2001. https://history.state.gov/historicaldocuments/frus1969–76v03/d160.

Kaufman, Henry. "'What Financial Change Has Wrought: The Last Half Century and Beyond,' An Address Before the National Economists Club." Washington, DC, October 17, 2018.

Klein, Herbert C. "Memorandum for the President from Herbert C. Klein, August 16, 1971." Box 68, Presidential Speech File, President's Personal Files, Richard M. Nixon Presidential Library and Museum, Yorba Linda, CA.

Kummer, Steve, and Christian Pauletto. "The History of Derivatives: A Few Milestones." Presented at the EFTA Seminar on Regulation of Derivatives Markets, Zurich, May 3, 2012.

"Letter from President Nixon to President Pompidou, February 16, 1972." Document 224 in *Foreign Relations of the United States, 1969–1976.* Vol. 3, *Foreign Economic Policy; International Monetary Policy, 1969–1972.* Washington, DC: U.S. Government Printing Office, 2001. https://history.state.gov/historicaldocuments/frus1969–76v03/d224.

"Letter from President Pompidou to President Nixon, February 4, 1972." Document 223 in *Foreign Relations of the United States, 1969–1976.* Vol. 3, *Foreign Economic Policy; International Monetary Policy, 1969–1972.* Washington, DC: U.S. Government Printing Office, 2001. https://history.state.gov/historicaldocuments/frus1969–76v03/d223.

McCracken, Paul. "Memorandum for the President," August 9, 1971. FRASER, https://fraser.stlouisfed.org/archival/1173#3382.

———. "Memorandum for the President, August 16, 1971." Richard M. Nixon Presidential Library and Museum, Yorba Linda, CA.

———. "Memorandum for the President, Subject: Quadriad Meeting— Monday, June 14, 1971," June 14, 1971. FRASER, https://fraser.stlouis fed.org/archival/1173/item/3391.

———. "Memorandum from the Chairman of the Council of Economic Advisers (McCracken) to President Nixon, June 2, 1971." Document 157 in *Foreign Relations of the United States, 1969–1976*. Vol. 3, *Foreign Economic Policy; International Monetary Policy, 1969–1972*. Washington, DC: U.S. Government Printing Office, 2001. https://history.state.gov /historicaldocuments/frus1969–76v03/d157.

"Memorandum from Secretary of the Treasury Kennedy to President Nixon, June 23, 1969 (Includes Paper, 'Basic Options in International Monetary Affairs,' by Paul Volcker)." Document 130 in *Foreign Relations of the United States, 1969–1976*. Vol. 3, *Foreign Economic Policy; International Monetary Policy, 1969–1972*. Washington, DC: U.S. Government Printing Office, 2001. https://history.state.gov/historicaldocuments /frus1969–76v03/d130.

"Memorandum from the President's Assistant for International Economic Affairs (Peterson) to President Nixon, September 20, 1971. Subject: Coordinating Group-Planning the Negotiations for the New Economic Policy Abroad." Document 177 in *Foreign Relations of the United States, 1969– 1976*. Vol. 3, *Foreign Economic Policy; International Monetary Policy, 1969–1972*. Washington, DC: U.S. Government Printing Office, 2001. https://history.state.gov/historicaldocuments/frus1969–76v03/d177.

"Memorandum of Conversation, Subject: President Nixon's New Economic Program, August 16, 1971." Document 170 in *Foreign Relations of the United States, 1969–1976*. Vol. 3, *Foreign Economic Policy; International Monetary Policy, 1969–1972*. Washington, DC: U.S. Government Printing Office, 2001. https://history.state.gov/historicaldocuments/ frus1969–76v03/d170.

"Memorandum of Conversation, Subject: President Nixon's New Economic Program, August 17, 1971." Document 171 in *Foreign Relations of the United States, 1969–1976*. Vol. 3, *Foreign Economic Policy; International Monetary Policy, 1969–1972*. Washington, DC: U.S. Government Printing Office, 2001. https://history.state.gov/historicaldocuments /frus1969–76v03/d171.

Mishkin, Frederic S., and Eugene N. White. "Unprecedented Actions: The Federal Reserve's Response to the Global Financial Crisis in Historical

Perspective." Working Paper, October 2014. https://www.dallasfed.org
/-/media/documents/institute/wpapers/2014/0209.pdf.

"Negotiating the New Economic Policy Abroad-Working Group," October
26, 1971. Richard M. Nixon Presidential Library and Museum. Yorba
Linda, CA.

"New Realities and New Directions in U.S. Foreign Economic Policy." Re-
port by the Subcommittee on Foreign Economic Policy of the House
Committee on Foreign Affairs, February 28, 1972. 92nd Cong., 1st Sess.
ProQuest Congressional.

Nixon, Richard M. "Address to the Congress on Stabilization of the Economy. September 9, 1971." Document 287 in *Public Papers of the Presidents of the United States: Richard Nixon, 1971*, 938–44. Washington,
DC: U.S. Government Printing Office, 1972.

———. "Address to the Nation Announcing Decision to Resign the Office of the President of the United States, August 8, 1974." American Presidency Project. https://www.presidency.ucsb.edu/documents
/address-the-nation-announcing-decision-resign-the-office-president
-the-united-states.

———. "Address to the Nation on the Rising Cost of Living, October 17,
1969." Document 395 in *Public Papers of the Presidents of the United
States: Richard Nixon, 1969*, 808–12. Washington, DC: U.S. Government Printing Office, 1971.

———. "Address to the Nation Outlining a New Economic Policy: The
Challenge of Peace." Document 264 in *Public Papers of the Presidents of
the United States: Richard Nixon, 1971*, 886–91. Washington, DC: U.S.
Government Printing Office, 1972.

———. "Annual Message to Congress on the State of the Union, January
22, 1970." Document 9 in *Public Papers of the Presidents of the United
States: Richard Nixon, 1970*, 8–16. Washington, DC: U.S. Government
Printing Office, 1971.

———. "The Challenge of Peace." President Nixon's Address to the Nation
on a New Economic Policy. https://youtu.be/ye4uRvkAPhA.

———. "First Annual Report to the Congress on United States Foreign Policy for the 1970s, February 18, 1970." Document 45 in *Public Papers of
the Presidents of the United States: Richard Nixon, 1970*, 116–190. Washington, DC: U.S. Government Printing Office, 1971.

———. "Inaugural Address." Document 1 in *Public Papers of the Presidents
of the United States: Richard Nixon, 1969*, 1–4. Washington, DC: U.S.
Government Printing Office, 1971.

———. "Informal Remarks in Guam with Newsmen, July 25, 1969." Document 279 in *Public Papers of the Presidents of the United States: Richard Nixon, 1969*, 549. Washington, DC: U.S. Government Printing Office, 1971.

———. *International Economic Report of the President, 1973*. Washington, DC: U.S. Government Printing Office, March 1973. http://hdl.handle.net/2027/msu.31293022471266.

———. *International Economic Report of the President, 1974*. Washington, DC: U.S. Government Printing Office, 1974. FRASER, https://fraser.stlouisfed.org/title/45/item/8145.

———. "President Nixon's Speech on the Economy." Aka "Nixon Shock Speech," with NBC commentary; ABC commentary featuring Paul McCracken; CBS commentary (8/15/1971). WHCA-4582, White House Communications Agency Videotape Collection, Richard Nixon Presidential Library and Museum, Yorba Linda, CA.

———. *Presidential Daily Diary, 1969*. White House Central Files, Staff Member and Office Files, Office of Presidential Papers and Archives, Richard M. Nixon Presidential Library and Museum. Yorba Linda, CA. https://www.nixonlibrary.gov/president/presidential-daily-diary.

———. *Presidential Daily Diary, 1971*. White House Central Files, Staff Member and Office Files, Office of Presidential Papers and Archives, Richard M. Nixon Presidential Library and Museum, Yorba Linda, CA. https://www.nixonlibrary.gov/president/presidential-daily-diary.

———. "The President's News Conference of August 4, 1971." Document 250 in *Public Papers of the Presidents of the United States: Richard Nixon, 1971*, 849–61. Washington, DC: U.S. Government Printing Office, 1971.

———. "The President's News Conference of February 6, 1969." Document 34 in *Public Papers of the Presidents of the United States: Richard Nixon, 1969*, 66–76. Washington, DC: U.S. Government Printing Office, 1971.

———. "The President's News Conference of January 27, 1969." Document 10 in *Public Papers of the Presidents of the United States: Richard Nixon, 1969*. Washington, DC: U.S. Government Printing Office, 1971.

———. "The President's News Conference of September 16, 1971." Document 292 in *Public Papers of the Presidents of the United States: Richard Nixon, 1971*, 949–59, Washington, DC: U.S. Government Printing Office, 1972.

———. "Remarks Announcing a Monetary Agreement Following a Meeting of the Group of Ten, December 18, 1971." Document 401 in *Public Papers of the Presidents of the United States: Richard Nixon, 1971*, 1195. Washington, DC: U.S. Government Printing Office, 1972.

————. "Remarks Announcing Appointment of Peter G. Peterson as Assistant to the President for International Economic Affairs and Executive Director, Council on International Economic Policy, January 19, 1971." Document 19 in *Public Papers of the Presidents of the United States: Richard Nixon, 1971*, 42. Washington, DC: U.S. Government Printing Office, 1972.

————. "Remarks at a Questions-and-Answer Session with a 10-Member Panel of the Economic Club of Detroit. September 23, 1971." Document 297 in *Public Papers of the Presidents of the United States: Richard Nixon, 1971*, 965–80. Washington, DC: U.S. Government Printing Office, 1972.

————. "Remarks at the Annual Meeting of the National Association of Manufacturers, December 4, 1970." Document 447 in *Public Papers of the Presidents of the United States: Richard Nixon, 1970*, 1088–95. Washington, DC: U.S. Government Printing Office, 1971.

————. "Remarks at the Swearing in of Dr. Arthur F. Burns as Chairman of the Board of Governors of the Federal Reserve System, January 31, 1970." Document 21 in *Public Papers of the Presidents of the United States: Richard Nixon, 1970*, 44–46. Washington, DC: U.S. Government Printing Office, 1971.

————. "Remarks on Plans to Nominate Secretary Kennedy as Ambassador-at-Large and Governor Connally as Secretary of the Treasury, December 14, 1970." Document 460 in *Public Papers of the Presidents of the United States: Richard Nixon, 1970*, 1129–31. Washington, DC: U.S. Government Printing Office, 1971.

————. "Remarks to Midwestern News Media Executives Attending a Briefing on Domestic Policy in Kansas City, Missouri, July 6, 1971." Document 222 in *Public Papers of the Presidents of the United States: Richard Nixon, 1971*, 802. Washington, DC: U.S. Government Printing Office, 1972.

————. "Remarks to Officials of the International Monetary Fund and the International Bank for Reconstruction and Development, September 29, 1971." Document 316 in *Public Papers of the Presidents of the United States: Richard Nixon, 1971*, 1014–16. Washington, DC: U.S. Government Printing Office, 1972.

————. "Remarks to Top Personnel at the Department of the Treasury, February 14, 1969." Document 49 in *Public Papers of the Presidents of the United States: Richard Nixon, 1969*, 101–5. Washington, DC: U.S. Government Printing Office, 1971.

———. "Second Annual Report to the Congress on United States Foreign Policy, February 25, 1971." Document 75 in *Public Papers of the Presidents of the United States: Richard Nixon, 1971*, 293. Washington, DC: U.S. Government Printing Office, 1972.

———. "Statement About Signing Par Value Modification Act, April 3, 1972." In *Public Papers of the Presidents of the United States: Richard Nixon, 1972*, 513–14. Washington, DC: U.S. Government Printing Office, 1974.

———. "Statement of the Balance of Payments, April 4, 1969." Document 141 in *Public Papers of the Presidents of the United States: Richard Nixon, 1969*, 265–66. Washington, DC: U.S. Government Printing Office, 1971.

———. "Telegram from the Department of State to the Embassy in Germany, August 16, 1971." Document 169 in *Foreign Relations of the United States, 1969–1976*. Vol. 3, *Foreign Economic Policy; International Monetary Policy, 1969–1972*. Washington, DC: U.S. Government Printing Office, 2001. https://history.state.gov/historicaldocuments /frus1969–76v03/d169.

———. "Transcripts of Acceptance Speeches by Nixon and Agnew to the G.O.P. Convention." *New York Times*, August 9, 1968.

Ohlmacher, Scott W. "The Dissolution of the Bretton Woods System Evidence from the Nixon Tapes, August–December 1971." Honors thesis, University of Delaware, 2009. http://udspace.udel.edu/handle/19716/4275.

Organisation for Economic Co-operation and Development. *The Future of Money*. Paris, France: OECD, 2002.

"Paper Prepared in the Department of the Treasury: 'Contingency,' May 8, 1971." Document 152 in *Foreign Relations of the United States, 1969–1976*. Vol. 3, *Foreign Economic Policy; International Monetary Policy, 1969–1972*. Washington, DC: U.S. Government Printing Office, 2001. https://history.state.gov/historicaldocuments/frus1969–76v03/d152.

"Paper Prepared in the Department of the Treasury: 'U.S. Negotiating Position on Gold,' May 9, 1971." Document 153 in *Foreign Relations of the United States, 1969–1976*. Vol. 3, *Foreign Economic Policy; International Monetary Policy, 1969–1972*, Washington, DC: U.S. Government Printing Office, 2001. https://history.state.gov/historicaldocuments/frus1969–76v03/d153.

Papers of Paul A. Volcker. Box 108477. Federal Reserve Bank of New York Archives.

Peterson, Peter G. "Memorandum for the President: Projecting the Future Development of the U.S.," July 12, 1971, "CIEP Folder." National Archives, College Park, MD.

———. "Memorandum from Peter Peterson to John Connally, 'Negotiating the New Economic Policy—Work Group,'" October 26, 1971. Folder: "Council on International Economic Policy." National Archives, College Park, MD.

———. "Negotiating the New Economic Policy Abroad," September 23, 1971. Richard M. Nixon Presidential Library and Museum, Yorba Linda, CA.

———. "Peter G. Peterson, Memorandum to Secretaries of State, Treasury, Director OMB, Chairman CEA, Chairman, Federal Reserve System, Assistant to the President for National Security Affairs, 'Negotiating the New Economic Policy Abroad,'" September 23, 1971. Folder: "Council on International Economic Policy." National Archives, College Park, MD.

———. "The United States in the Changing World Economy." Washington, DC: U.S. Government Printing Office, 1971.

———. "The United States in the Changing World Economy (Confidential Version)." April 1971. Box 108477. Papers of Paul A. Volcker, New York Federal Reserve Archive.

"Remarks of the President, Secretary of the Treasury John B. Connally, and Secretary of State William P. Rogers on Arrival at Andrews Air Force Base, December 14, 1971." In *Weekly Compilation of Presidential Documents* 7, no. 51 (December 20, 1971): 1665–66. https://hdl.handle.net/2027/osu.32437011290398?urlappend=%3Bseq=722.

"Reshaping the International Economic Order." Washington, DC: Brookings Institution, January 1972. ProQuest.

Ribicoff, Abraham. *Trade Policies in the 1970s: Report by Senator Abraham Ribicoff to the Committee on Finance, United States Senate*. S. Pt. 92–1. Washington, DC: U.S. Government Printing Office, 1971. https://www.finance.senate.gov/imo/media/doc/Sprt2.pdf.

Ritter, Christen Thomas. "Closing the Gold Window: Gold, Dollars, and the Making of Nixonian Foreign Economic Policy." PhD diss., University of Pennsylvania, 2007.

Shafer, Jeffrey R., and Bonnie E. Loopesko. "Floating Exchange Rates after Ten Years." *Brookings Papers on Economic Activity* 1983, no. 1 (1983): 1–86. https://doi.org/10.2307/2534352.

Shultz, George P. "Prescription for Economic Policy: 'Steady As You Go,' Address Before the Economic Club of Chicago, April 22, 1971." https://web.stanford.edu/~johntayl/Shultz%20on%20Steady%20As%20You%20Go.pdf.

Stabler, Elizabeth. "The Dollar Devaluations of 1971 and 1973." In *Commission on the Organization of the Government for the Conduct of Foreign Policy: Appendices*, 3:139–59. Washington, DC, 1975. http://hdl.handle.net/2027/ucl.31158010972742.

Symington, Stuart. (Statement of Senator Symington.) "Further Concentration of Power, Executive Privilege, and the 'Kissinger Syndrome.'" In *Congressional Record*. Vol. 117, Part 4:4498–503. 92nd Cong., 1st Sess. (1971). ProQuest Congressional.

"Text of Group of Ten Communiqué." *Department of State Bulletin* 66, no. 1698 (January 10, 1972): 32–34. https://hdl.handle.net/2027/mdp.39015077199571?urlappend=%3Bseq=38.

"Text of Joint Statement (in Azores), December 14, 1971." *Department of State Bulletin* 66, no. 1698 (January 10, 1972): 30–31. https://hdl.handle.net/2027/mdp.39015077199571?urlappend=%3Bseq=36.

"Trade (% of GDP)—Germany | Data." Chart. https://data.worldbank.org/indicator/NE.TRD.GNFS.ZS?end=2018&locations=DE&start=1960&view=chart.

"Trade (% of GDP)—Japan | Data." Chart. https://data.worldbank.org/indicator/NE.TRD.GNFS.ZS?end=2018&locations=JP&start=1960&view=chart.

Transcript of Richard Nixon tapes. Tape 741–002, June 23, 1972, Oval Office, 10:04–11:39. Richard M. Nixon Presidential Library and Museum, Yorba Linda, CA. https://www.nixonlibrary.gov/sites/default/files/forresearchers/find/tapes/watergate/wspf/741–002.pdf.

"Treasury Meeting with Academic Consultants, 'Summary of Afternoon Discussion on International Matters,'" August 25, 1971. National Archives, College Park, MD.

United States Commission on International Trade and Investment Policy (Williams Commission). *United States International Economic Policy in an Interdependent World; Report to the President (The Williams Commission Report)*. Washington, DC: U.S. Government Printing Office, 1971. https://catalog.hathitrust.org/Record/001122081.

United States Congress. *Hearing Before the Committee on Labor and Public Welfare on George P. Shultz to be Secretary of Labor* (1969). 91st Cong., 1st Sess. (January 16, 1969). ProQuest Congressional.

———. *Nomination of Arthur F. Burns: Hearing Before the Committee on Banking and Currency, United States Senate.* 91st Cong. 1st Sess. (December 18, 1969). https://fraser.stlouisfed.org/scribd/?title_id=783&fil epath=/files/docs/historical/senate/burns_confirmation.pdf.

———. *The President's New Economic Program: Hearings Before the Joint Economic Committee.* 92nd Cong., Part 1 (1971). ProQuest Congressional.

———. *The President's New Economic Program: Hearings Before the Joint Economic Committee,* 92nd Cong., Part 2 (1971). ProQuest Congressional

———. *The President's New Economic Program: Hearings Before the Joint Economic Committee,* 92nd Cong., Part 3 (1971). ProQuest Congressional

———. *The President's New Economic Program: Hearings before the Joint Economic Committee,* 92nd Cong., Part 4 (1971). ProQuest Congressional

———. "Report of the Joint Economic Committee, Congress of the United States on the February 1971 Economic Report of the President," March 30, 1971. Washington, DC: U.S. Government Printing Office, 1971. https://www.jec.senate.gov/reports/92nd%20Congress/Joint%20Economic%20Report%20on%20the%201971%20Economic%20Report%20of%20the%20President%20(507).pdf.

United States Department of Commerce, Bureau of the Census. *Fourteenth Census of the United States Taken in the Year 1920.* Vol. 1, *Population, 1920. Number and Distribution of Inhabitants.* Washington, DC: U.S. Government Printing Office, 1921. https://www.census.gov/prod/www/decennial.html.

———. *Thirteenth Census of the United States Taken in the Year 1910.* Vol. 2, *Population 1910: Reports by States Alabama–Montana.* Washington, DC: U.S. Government Printing Office, 1913. https://www.census.gov/prod/www/decennial.html#y1910fin.

United States Department of State. *Foreign Relations of the United States, 1969–1976.* Vol. 3, *Foreign Economic Policy; International Monetary Policy, 1969–1972.* Edited by Bruce F. Duncombe and David S. Patterson. Washington, DC: U.S. Government Printing Office, 2001. https://history.state.gov/historicaldocuments/frus1969–76v03.

United States Department of the Treasury. "Report to the Congress of the Commission on the Role of Gold in the Domestic and International Monetary Systems." Washington, DC, March 1982. FRASER https://fraser.stlouisfed.org/title/339.

———. "Treasury Designates China as a Currency Manipulator," August 5, 2019. https://home.treasury.gov/index.php/news/press-releases/sm751.

————. "Treasury Releases Report on Macroeconomic and Foreign Exchange Policies of Major Trading Partners of the United States," May 28, 2019. https://home.treasury.gov/index.php/news/press-releases/sm696.

United States President and Council of Economic Advisers. *Economic Report of the President: Transmitted to Congress Together with the Annual Report of the Council of Economic Advisers (1969)*. Washington, DC: U.S. Government Printing Office, 1969. FRASER, https://fraser.stlouisfed.org/title/45#8140.

————. *Economic Report of the President: Transmitted to Congress Together with the Annual Report of the Council of Economic Advisers (1970)*. Washington, DC: U.S. Government Printing Office, 1970. FRASER, https://fraser.stlouisfed.org/title/economic-report-president-45?browse=1950s#8141.

————. *Economic Report of the President: Transmitted to Congress Together with the Annual Report of the Council of Economic Advisers (1971)*. Washington, DC: U.S. Government Printing Office, 1971. FRASER, https://fraser.stlouisfed.org/title/45#8142.

————. *Economic Report of the President: Transmitted to Congress Together with the Annual Report of the Council of Economic Advisers (1972)*. Washington, DC: U.S. Government Printing Office, 1972. FRASER, https://fraser.stlouisfed.org/title/45#8143.

————. *Economic Report of the President: Transmitted to Congress Together with the Annual Report of the Council of Economic Advisers (1973)*. Washington, DC: U.S. Government Printing Office, 1973. FRASER, https://fraser.stlouisfed.org/title/45#8144.

————. *Economic Report of the President: Transmitted to Congress Together with the Annual Report of the Council of Economic Advisers (1974)*. Washington, DC: U.S. Government Printing Office, 1974. FRASER, https://fraser.stlouisfed.org/title/45#8145.

"Urgent Information Memorandum from Robert Hormats and Helmut Sonnenfeldt of the National Security Council Staff to the President's Assistant for National Security Affairs (Kissinger), January 24, 1972." Document 222 in *Foreign Relations of the United States, 1969–1976*. Vol. 3, *Foreign Economic Policy; International Monetary Policy, 1969–197*. Washington, DC: U.S. Government Printing Office, 2001. https://history.state.gov/historicaldocuments/frus1969–76v03/d222.

"U.S. Gold Stock and World Monetary Gold Holdings," August 15, 1971. FRC Box 7. Record Group 56. Records of Secretary of the Treasury George P. Shultz, 1971–1974. National Archives, College Park, MD.

Various documents. Box 175. Staff Member Office Files, Alphabetical Subject Files, Camp David, Richard M. Nixon Presidential Library and Museum, Yorba Linda, CA.

Various documents. FRC Box 7. Record Group 56. Records of Secretary of the Treasury George P. Shultz, 1971–1974. National Archives, College Park, MD.

Volcker, Paul A. Interview with Paul A. Volcker. January 28, 2008–March 24, 2010. Federal Reserve Board Oral History Project. https://www.federal reserve.gov/aboutthefed/files/paul-a-volcker-interview-20080225.pdf.

———. Interview with Paul Volcker. *Commanding Heights*. PBS, September 26, 2000. http://www.pbs.org/wgbh/commandingheights/shared /minitextlo/int_paulvolcker.html.

———. "Remarks by Paul A. Volcker at the Bretton Woods Committee Annual Meeting 2014." The Bretton Woods Committee, May 21, 2014. https://www.brettonwoods.org/publication/remarks-by-paul-a-volcker -at-the-bretton-woods-committee-annual-meeting-2014.

———. "Telegram from the Embassy in the United Kingdom to the Department of State, September 17, 1971." Document 175 in *Foreign Relations of the United States, 1969–1976*. Vol. 3, *Foreign Economic Policy; International Monetary Policy, 1969–1972*. Washington, DC: U.S. Government Printing Office, 2001. https://history.state.gov/historicaldocuments/ frus1969–76v03/d175.

"The White House: Explanatory Material on President's Economic Program," August 15, 1971. FRC Box 7. Record Group 56. Records of Secretary of the Treasury George P. Shultz, 1971–1974. National Archives, College Park, MD.

"White House Photo Collection." Book 31. n.d. Richard M. Nixon Presidential Library and Museum, Yorba Linda, CA.

World Economic Forum. "Central Bank Digital Currency Policy-Maker Toolkit." January 22, 2020. White Papers. World Economic Forum. https://www.weforum.org/whitepapers/central-bank-digital-currency -policy-maker-toolkit.

Author Interviews

Baker, James. Former secretary of state and former secretary of the treasury. Interviewed November 28, 2017.

Balladur, Édouard. Former chief adviser to George Pompidou, president of France. Interviewed February 8, 2018.

Barnes, Ben. Former state Speaker of the House in Texas (under John Connally as governor). Interviewed January 12, 2018.

Bergsten, C. Fred. Former staff member of the National Security Council and informal consultant to Henry Kissinger. Interviewed October 5, 2017.

Bradfield, Michael. Former Treasury deputy counsel; present at Camp David. Interviewed July 12, 2017.

Cassidy, Robert. Former staff, Senate Finance Committee. Interviewed October 18, 2017.

Cooper, Richard. Former professor at Yale University and consultant to Henry Kissinger. Interviewed June 7, 2018.

Cummings, Frank. Former staff member, Sen. Jacob Javits. Interviewed December 6, 2017.

Dam, Kenneth. Former associate director of the Office of Management and Budget; present at Camp David. Interviewed July 18, 2017.

Daniels, Mike. Attorney for Japanese textile interests in late 1960s and early '70s. Interviewed October 5 and 17, 2017.

D'Arista, Jane. Former staff member, Joint Economic Committee. Interviewed October 12, 2017.

Galbraith, James. Former staff member, Joint Economic Committee; former staff member, Congressman Henry Reuss. Interviewed November 27, 2017.

Giscard d'Estaing, Valéry. Former finance minister and former president of France. Interviewed November 21, 2017.

Greenspan, Alan. Former chairman, Council of Economic Advisers; former chairman, Federal Reserve. Interviewed November 15, 2017.

Guenther, Ken. Former staff member, Sen. Jacob Javits. Interviewed December 6, 2017.

Hormats, Robert D. Former staff member, National Security Council. Interviewed July 12 and December 11, 2017.

Huebner, Lee. Former speechwriter for Richard Nixon. Interviewed November 15, 2017, and February 1, 2018.

Issing, Otmar. Long-term German expert on finance. Interviewed February 6, 2018.

Jasinowski, Jerry. Former staff member, Joint Economic Committee. Interviewed October 5, 2017.

Kaufman, Henry. Former senior partner, Solomon Brothers; current president of Henry Kaufman & Co. Interviewed July 6, 2017.

Kissinger, Henry A. National security advisor to Richard Nixon and former secretary of state. Interviewed July 10, 2018.

Koed, Betty. Senate historian. Interviewed September 20, 2017.

Kohn, Don. Former vice chairman, Federal Reserve System. Interviewed October 18, 2017.

Korologos, Tom. Former congressional staff, Richard Nixon. Interviewed October 6, 2017.

Levinson, Marc. Economist and author of *An Extraordinary Time*. Interviewed November 9, 2017.

Levy, Charles. Former staff, Senate Foreign Relations Committee. Interviewed October 11, 2017.

Lissakers, Karin. Former staff director, Senate Subcommittee on Foreign Economic Policy. Interviewed October 19, 2017.

Mayer, Martin. Author of *The Fate of the Dollar*. Interviewed October 19, 2017.

McLean, Ken. Staff member, Sen. William Proxmire. Interviewed October 6, 2017.

Morris, Dennis. U.S. Marine guard, Camp David, 1970. Interviewed January 19, 2018.

Mulford, David. Former undersecretary of the treasury. Interviewed December 5, 2017.

Nolan, Charles. U.S. Marine guard, Camp David, 1970. Interviewed January 19, 2018.

Petty, John. Former assistant secretary of the treasury for international affairs, Nixon administration. Interviewed October 11 and November 9, 2017.

Read, Julian. Former press secretary to Governor Connally. Interviewed January 24, 2018.

Rubin, Robert. Former secretary of the treasury and former co-chairman of Goldman Sachs. Interviewed October 31, 2017.

Schlesinger, Helmut. Former president, German Central Bank (Bundesbank). Interviewed by email exchanges. April 30, 2018.

Shultz, George P. Former secretary of labor; director OMB; secretary of the treasury; and secretary of state; present at Camp David. Interviewed August 16, 2017.

Steil, Benn. Economist and author of *The Battle of Bretton Woods: John Maynard Keynes, Harry Dexter White, and the Making of a New World Order*, and *The Marshall Plan: Dawn of the Cold War*. Interviewed July 26, 2017.

Temple, Larry. Former chief of staff, Governor Connally. Interviewed January 24, 2018.

Truman, Edwin. Former assistant secretary of the Treasury; former director, Division of International Finance, Federal Reserve System. Interviewed June 28, 2017.

Volcker, Paul A. Former undersecretary of the treasury for monetary affairs; president, New York Federal Reserve; chairman, Federal Reserve; present at Camp David. Interviewed July 26, 2017.

Weber, Arnold. Former deputy director, Office of Management and Budget; present at Camp David. Interviewed August 14, 2017.

Whitman, Marina von Neumann. Former staff member and then member, Council of Economic Advisers. Interviewed December 2, 2017.

Wolff, Alan. Former counsel, Treasury Department. Interviewed December 14, 2018.

INDEX

ABOUT THE AUTHOR

Jeffrey E. Garten is dean emeritus of the Yale School of Management, where he teaches courses on the global economy. Formerly he was undersecretary of commerce for international trade in the Clinton administration, and a managing director of the Blackstone Group. He is the author of *From Silk to Silicon* and several other books on economics, business, and foreign policy.

READ MORE BY JEFFREY GARTEN

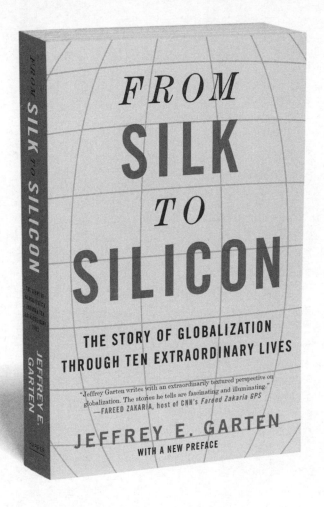

"This is a tale of globalization and leadership that is both sweeping and personal. By focusing on ten transformational people, it shows how individuals can affect the flow of history. It's a guide to the future as well as to the past."

—WALTER ISAACSON,
author of *Steve Jobs, Einstein,* and *Benjamin Franklin*